JOHN MITCHELL

John Mitchell. 1900

JOHN MITCHELL

Miner

LABOR'S BARGAIN WITH THE GILDED AGE

by Elsie Glück

GREENWOOD PRESS, PUBLISHERS
NEW YORK

Reprinted by permission
of the John Day Company

First Greenwood Reprinting 1969

Library of Congress Catalogue Card Number 69-13909

SBN 8371-2170-1

PRINTED IN UNITED STATES OF AMERICA

To

MY MOTHER

ABOUT THE AUTHOR

ELSIE GLÜCK has devoted many years to the study of labor problems, and through her work with trade unions has first-hand knowledge.

After being graduated from the University of Wisconsin in 1920, where she was a student of John R. Commons, she directed the educational work of the Ladies' Waist and Dressmakers' Union. Subsequently, as a member of the staff of the International Ladies' Garment Workers' Union, she conducted campaigns and assisted in strikes in several cities. She has also done research work for that organization, after having served as research assistant to several well-known economists. She has been more recently connected with the teaching staff of the Economics Department of the University of Wisconsin.

From time to time, she has contributed articles to the *American Federationist*, *Advance* and other labor publications.

CONTENTS

LIST OF ILLUSTRATIONS

ACKNOWLEDGMENTS

The author wishes to acknowledge indebtedness to Miss Elizabeth C. Morris, secretary to John Mitchell, for her assistance in making available the many documents of mine workers' conventions, joint conferences, etc., as well as for her many suggestions and corrections during the writing of this book. The early files of the *United Mine Workers' Journal* were made accessible through the courtesy of Mr. Thomas Kennedy, Secretary-Treasurer of the United Mine Workers' International Union. This source material was supplemented by contemporary newspaper and periodical comment, and the author also drew heavily on *Conciliation and Arbitration in the Coal Industry of America* by Arthur E. Suffern.

In the effort to supplement by personal interview information obtainable from published sources, the author interviewed dozens of persons connected with Mr. Mitchell. She wishes to acknowledge with special thanks the assistance of Miss Morris and of Mr. W. D. Ryan, formerly secretary-treasurer of the United Mine Workers' organization, now with the U. S. Bureau of Mines. Among the many others who have generously contributed of their time are:

Mr. Richard J. Beamish, formerly with the *Philadelphia North American;*
Father John J. Curran of Wilkesbarre, Pa.;
Mr. Clarence S. Darrow, Chicago;

Mr. Thomas Duffy, formerly President of Anthracite District, No. 7 (U.M.W.);

Mr. Thomas Duffy, newspaper correspondent, Scranton, Pa.;

Mrs. Mary Burke East, official stenographer for all mine workers' conventions and conferences, as well as for the conventions of the A. F. of L.;

Mr. Frank Feehan, formerly President of Pittsburgh District, No. 5, U.M.W.A.;

Mr. W. H. Haskins, formerly President of Ohio District, No. 6;

Mr. Daniel Hart, now Mayor of Wilkesbarre, Pa.;

Mr. Thomas Kennedy, National Secretary-Treasurer of Mine Workers' Union;

Mr. John Loftus, Scranton;

Mr. Andrew Matti, President of Anthracite District, No. 7;

Mr. Duncan McDonald, at one time on the staff of the Union;

Mr. James O'Connell, Metal Trades Department, American Federation of Labor;

Mr. Charles P. Neill, Umpire under agreement in anthracite fields;

Mr. Michael D. Ratchford, President of U.M.W., 1897 and 1898;

Miss Harriett Reid, formerly of the staff of the U.M.W.A.;

Mr. Harry N. Taylor, formerly an operator in Illinois;

Mr. W. D. Van Horn, long a National Executive Board member of U.M.W.A.;

Mr. John H. Walker of Springfield, Ill., one-time officer of U.M.W.A., now President of Illinois State Federation of Labor;

Dr. Frank Julian Warne, economist, of Washington;

Mr. Walter Wellman, one-time correspondent for *The Chicago Record-Herald;*

Mrs. Walter Weyl, whose husband was associated with John Mitchell, and who collaborated with him in the writing of *Organized Labor.*

It is to be regretted that the official correspondence of Mr. Mitchell during his years of office was unavailable. These records were reported as missing when John P. White entered office in 1911. Through the courtesy of President William

Green and Miss Florence C. Thorne of the American Federation of Labor, the correspondence of Mr. Mitchell and Samuel Gompers was made available. Matters covered by personal correspondence and personal files are contemplated for use in a volume by Miss Morris and are therefore not accessible. Personal correspondence with John Loftus of Scranton, Pa., was placed at my disposal by Mr. Loftus. Mrs. Caro Lloyd also permitted me to examine such correspondence of Henry D. Lloyd as might throw light on the anthracite strike of 1902.

Research in connection with this study was carried on while the author was a fellow at the Robert Brookings Graduate School of Economics and Government, and as a graduate student at the University of Wisconsin. The visits to mining centers were made possible through the aid of the American Fund for Personal Service.

Criticisms and suggestions on the manuscript were made by Mr. Arthur E. Suffern, now of the Federal Council of Churches; Mr. W. D. Ryan, Dr. Leo Wolman, Director of Research for the Amalgamated Clothing Workers' Union, by Miss Marguerite Steinfeld and Miss Marjorie Griesser. The first half of the original draft was also carefully read and criticized by Dr. Carter Goodrich of the University of Michigan, Dr. I. Lubin and Dr. Helen R. Wright of the Institute of Economics. Mr. H. S. Raushenbush gave invaluable aid in the revision of the manuscript. I am especially indebted to Professors John R. Commons and Selig Perlman of the University of Wisconsin for such counsel and continual encouragement as made possible the completion of this study.

INTRODUCTION

THIS book is more than a biography—it is a study of leadership and mass movement. Miss Glück went where Mitchell went, and interviewed those who worked with or against Mitchell. His leadership had its individuality. It was created, as it were, out of the thin-vein coal deposits of Northern Illinois—the economist's "marginal mines"—where twice or three times as much work was required to get out a ton of coal as in the ten-foot veins of Southern Illinois and West Virginia. These were the last mines to open up and the first to shut down. Mitchell knew by experience—not by literature. Those veins produced many other thinkers, and talkers and leaders. But something about John Mitchell brought him to the leadership of the leaders. He was not aggressive—he was conciliatory. He talked after the others had finished. And it seemed that he could clearly state everything that had been said. You felt, too, a great power and integrity in his statements. Then, when he was quite certain that all were with him, he could talk in private or public to the organized employers with not only the same power and integrity but also with the still voice of 300,000. So he could wait. He was the least hurried of labor leaders. He let the operators have their say first, and then, as one of them said, he made them look like thirty cents. He did not exaggerate himself. He knew that he was a leader, not a boss. And he was just as philosophical when, flushed with power, the 300,000 would not wait for his conciliatory waiting. So he was a leader only a few years. A leader out of bondage—not a dictator after victory. Yet he left an enduring organization because he built on economic power to make collective agreements and on fidelity to them when disadvantageous. This biography is a useful lesson in leadership and democracy.

JOHN R. COMMONS.

University of Wisconsin.

JOHN MITCHELL

I

SONS OF TOIL AND SORROW

THE great burst of energy and ambition which ripped open the West and gave us some of our first and most notable industrial millionaires and buccaneers also shored up at the side of the railroads and coal mines some hundreds of dispirited, discouraged towns. The people in them had come from the East, had come all the way from England, from Scotland and Wales, to take their part in the opening of the industrial resources of the middle and western borders. Some of them went into Northern Illinois when the Central Illinois Railroad was built and settled in Decatur, Braceville, Braidwood and the small mining camps about. There was work for a while. Some of them went off to the Civil War. There was work for those who stayed at home. It was a country where there was a promise behind the hard work of pioneering, a time when men rose suddenly to wealth and power.

When the men came home from the war to their families even the promise of steady work faded, and the unspoken hopes beyond that essential were slowly buried by the process of keeping alive. There were more mines than the country needed. Many of them closed. The rest lowered wages. The miners and their families were left to shift for themselves.

The war had kept a united country—there had never been a war for a united coal industry. Competition was still seen as the life-blood of trade. What the miners suffered was a peaceful violence: unemployment, starvation wages. They had the right and freedom to try their fortune elsewhere in the growing country. Some of them even had the money to move away.

Habit and hope were strong and hardy in most of them. It was the mining craft they knew and wanted to work at. The closed collieries represented thousands of dollars of investment. They had worked once; surely they would work again if only men waited. Meanwhile some garden truck could be raised, some milk gotten from a cow or goat, some lucky chance would bring a few days' work on the roads or farms. Children kept coming into the world.

In the winter of 1870 John Mitchell was born in Braidwood, Illinois. His father had come back from the wars and stayed there to take his luck with the rest. His father, Robert Mitchell, was one of the Scotch-Irish who took a pride in being considered Irishmen at a time when their social standing in this country was low; but on July 12th he would be found marching in the Orangemen's Parade. He had three wives. His first wife gave him three sons before she died; his second gave him two, one of whom was John Mitchell, and a daughter, and then died when John was two and a half years old; his third wife brought along a child of her own from another marriage and gave John's father another son of his own. He was killed by a runaway team when John was six years old, and left these four sets of children under one roof and a heavy soldier's coat under which John and a half-brother would snuggle through the long cold winter nights.

The lives of the poor leave few documents. They are known to their neighbors and remembered only by the existence of something unusual. Poverty, a widow with four sets of children were not unusual. Of John's own mother only her name, Martha Halley, and her son are remembered. Of his stepmother there is such account as he himself gave in the later years when people sought to account for his career. Reluctantly he broke through the silence he had guarded about his childhood years.

His stepmother, a Scotch Presbyterian, had to take in washing. To her family she taught a few simple maxims. "Tell the truth and shame the devil" was one of them. She did not spoil her children. John had to help with the washing, and could attend school only irregularly. He could not keep pace with the other boys because of that. "The humiliation and shame of lagging behind them caused me to lose interest in my studies," he said later. This was only part of it. The other boys, whose mothers did not take in washing, would pass the house on Saturdays and Sundays to go to the ballgrounds, to the swimming-hole, to the woods. John had to remain at home, an exile from the life of the others, doing the washing, helping with the housework, weeding the vegetable garden. There was no relief on Sundays. Under the strict Presbyterian rule of his mother he went to Sunday School and read the Bible. Membership in the church counted for something to a woman who had little else to support her.

The boys of the town missed no chance to drive the exile

from their ranks further into his loneliness. When he hung the clothes out to dry in the backyard, they would gather around and torment him with the jibe that he was doing a girl's work. All this happened before he was ten years old but he never forgot it.

A normal boy could stand this only so long. When it was whispered around that the boys were going over to Joliet, twenty-one miles away, to see the circus, he took his courage in both hands and joined them. He had never been near one before. The parade was a scene that met his wildest expectations. The walk over had been long. None of them had any money for the fifty-cent ticket of admission. They were given the privilege of carrying water for the menagerie to earn their right to see the show. Without breakfast, without sleep all night during the walk, the boys worked at it seven hours, bucket after bucket. Finally the job was ended, the price of admission paid off. On the last trip, fifty yards away from the entrance, a boy of his own age approached him "and offered, most kindly, I thought, to carry my bucket for a short distance." He gladly surrendered it, only to see the good Samaritan walk right into the tent with what turned out to be the credential of the day's work, the last bucket. He never saw the circus. His childish belief in others, which was to last far beyond childhood, only brought him the experience of being seized by the collar and thrown out. On top of that he had the bitter return over the whole twenty-one miles, the taunts of his companions, and a beating from his stepmother when he reached home.

His stepmother married again when John was ten, and he

left home, wandering out into the neighboring countryside, finally getting a job on one of the farms as water boy to the farmhands. In addition to his keep he received one dollar a week. When he was twelve he came back to Braidwood and went into the mines. The State law put the age limit at thirteen years, but boys younger than himself were getting in. He had no great trouble getting a job as a trapper-boy, opening and closing the wooden doors that separated one section of the workings from another whenever the mule drivers came along the dark tracks with their coal trucks.

The pay was low. The day's work was long. He was one of thousands of boys who had come out of the loins of toil and sorrow that generated in the early seventies and who knew no better world. The men with whom he was later to fight his way up were going through the same life in the camps around him, in Indiana, Pennsylvania, Ohio. It was a time when men went to church because they had to believe in another world.

Braidwood and the other mining camps of the day were still places where the raw and simple function of emptying the bowels of the earth had not been embroidered by many of the delicacies of our culture. The tipple rose huge and unpainted above the miners' shacks that grouped around it. At one side was the growing mountain of culm, of waste. The places where men ate and slept were ramshackle affairs of one or two rooms. There was a general store, owned in most cases by the coal company, where the miners could buy food on the credit of each day's wage; a superintendent's house, a schoolhouse, many saloons and churches finished the picture. If a man wanted recreation he had the saloon at his door. In the periods of idleness he could try the rivers and woods for fish and game. The miners stayed by themselves. Some slight contact with the superintendent, the storekeeper, the saloonkeeper, priest or minister was the only connection they had with other walks of life, other manners of living.

The mining camp was isolated from other communities and modes of life, and the miner was isolated from other men. The entire social, industrial and political life centered about the mine. Such town officials as there were would most likely be miners, and the taste of local political life made many a miner ambitious to sit in the State assemblies and make better laws for the protection of his craftsmen. In John Mitchell's day, the mayor of Braidwood was one of the earliest of union leaders in Illinois—John James. Dan McLaughlin, who first attempted to put bargaining in the coal fields on a national scale was another of Braidwood's mayors.

Life in a mining camp was dreary but violent. Frequent strikes, the desperate outbursts of hungry men and boys, and race riots made these camps "tuff towns." Mine disasters on a large scale were comparatively infrequent, but death underground was a daily occurrence. The camp hardly knew a day without a funeral procession. Every home accepted the fact that its head might any day be brought home, injured or dead. The Diamond Mine Disaster of 1883 in Braceville, an adjoining town, resulted in the death of sixty-one men. A winter wind from the prairies froze the flooded mine. The miners came from the surrounding towns to help in the rescue, and John Walker of Illinois remembers seeing, in that procession, a trail of blood across the ice from under young Mitchell's cut bare feet.

The poverty that now seemed to have no end in sight, the constant risk of the occupation, drove men to drink. Still a little aside from his fellows, John Mitchell, at that time, would not drink. For this he was promptly and frequently charged by the young men of his age with being a "sissy." It was one of the worst taunts of boyhood, and he withstood it. The enforced difference that had so long separated him from the life of the other boys had become a virtue to which he was forced to cling tight. If he gave that up, he could only judge himself by the standards of the others, adult boys infinitely more shrewd than himself in the ways of the world.

None of these shrewd boys nor their fathers could really see any way out of the situation except the same process they used in leaving the lower levels of the workings for the top. The only way to get out of their situation was to get out of the mines. There was talk about unionism. There was always talk about that. But meanwhile they lived on as best they could. In Ohio annual earnings for miners, which were $314 in 1879 were dropping toward $239 in 1885 and 1886. For that pay men spent from ten to twelve hours underground in an isolation different only in degree from that they met aboveground.

The miner worked in a "room" or "breast" alone, unsupervised, on his own responsibility, shooting blasts, timbering the rock for safety, laying the track for the truck which was to carry out the coal, hauling water from his room, doing all the "dead work" for which he got no payment, incidental and necessary to the actual picking and loading of the coal. Sometimes the miner worked with a "buddy"; most often he worked alone, lying on his belly in the three-foot veins of Northern Illinois, bending over double as he loaded the coal, driving his mule in the darkness down the long passageways to the mine opening. One can visualize young John Mitchell at the age of twelve, going down to his first day in the mine, stepping on to the "cage" which bore men and coal up and down the shaft, being lowered several hundred feet into utter darkness, the feeble light in his top-heavy miner's cap blown out by the sudden gust of air. He recalls that he often slid down the rope before the arrival of the hoisting engineer, hours before the sunrise. At the bottom of the mine there were here and there faint lights on the caps of the men who hoisted cars of coal on to the platform. Then, probably, he was taken some distance, perhaps a mile, along the dark passages and left alone in the terrific darkness and silence, to open and shut the trapdoors which controlled the ventilation of the mine, in order to let cars go through, hearing only the strange noises of the mine—a driver swearing at his mule, the sound of a pick somewhere in the distance—and then silence, and waiting again.

Within a month after his entry into the mines, relief from this utter isolation came in the form of a promotion. The rise of a miner is from trapper-boy to mule driver, to miner's helper, to skilled miner, and then—when old age has set in—back again to the task of trapper-boy. Young Mitchell became helper to his stepfather, Smith, of whom he said in later years that he was a harsh teacher, but a good miner. These two would work away at their "room" in Northern Illinois. In low, three-foot veins, this meant a continual bending and bowing, a continual crouching, so that the miners of this region—according to the report of a State Investigating Commission for Illinois—came to have what was known as a "miner's shape." They developed a set of muscles "not developed by ordinary labor, at the expense of other muscles used in agricultural and everyday pursuits." This often unfitted them for other occupations.

To make matters worse, another troop in the long procession of immigrants was being brought into the coal fields. The first three British groups that had come—Scotch, Welsh and English—had had no love for each other. But they had joined against the Irish Catholic invaders with a hatred that had its roots in the days when the coal operators of Durham and Northumberland had replaced their striking miners with bewildered Irishmen who then knew nothing of the craft. Riots still occurred on St. Patrick's Day and Orangemen's Day. The Irish were not people to take a licking. They struck back. In the anthracite fields the Molly Maguires had gone out for the Welsh foremen. Now all four groups, coupled with some stray Germans, had to watch the importation of men they speedily called "roundheads"—Bohemian, Polish, Belgian and Italian workers. They brought with them a standard of living, a willingness "to work for a crust of bread and a kick in the pants," that threatened even the few dollars a week the other groups were getting out of the industry. Stories started, and have never been silenced, about some of these newcomers not only underbidding the others for the few jobs but offering to throw their wives into the bargain.

Some efforts were being made to meet the whole situation by a union of all the mine workers. The British miners had brought that tradition with them. One of the most active men lived right in Braidwood—Daniel McLaughlin—a follower of the great Scotch mine leader, Alex McDonald. Without much communication between the scattered camps, without resources, the task of binding together the men already in the mines against the operators or against the newcomers was almost superhuman. Two separate unions, the National Federation of Miners and the Knights of Labor, soon warring with each other, did come into the fields.

For a few years around 1885-6 it seemed possible that some less casual status for the mine workers might be established. In the country at large the Knights of Labor claimed seven hundred thousand members. The young American Federation of Labor claimed two hundred and fifty thousand. Prices had risen from 1880 to 1883. The exhaustion of the frontier land, the inpouring of cheap contract labor, a renewal of industrial expansion, all helped to make the workers see that they were being permanently shut up in the wage system. A series of strikes and lockouts began, some of them lasting over a year. In the Hocking Valley of Ohio the miners took a bad beating. But farmers and workers alike were beginning to have enough of "monopoly." A hatred of it lasted across lost strikes and high prices into 1885-6 and swelled the ranks of the Knights of Labor from various classes. A big victory in 1885 on the

Gould railroad showed that something could be done by organization.

It was this organization which John Mitchell and every other mine worker in Braidwood joined. Its romantic title had originally been backed up with pompous, elaborate initiations dwelling on the dignity of labor, the common tie between all classes. Although its primary appeal was to unskilled workers, in other parts of the country businessmen were members. It was to many miners an artificial leveler, as the church was to their wives. There was nothing revolutionary about it. In the pledge of membership he took, John Mitchell said with the others: "We mean no conflict with necessary enterprise, no antagonism to necessary capital." Its leaders were really more interested in education, mutual aid and labor legislation than in trade unionism.

The somewhat modest aims of this group, the appeal to liberal public opinion which it represented came to ruin in part by the fact that other groups in the country wanted much more and wanted it at once. The psychology of the disinherited was represented in the growth of the anarchist movement. The Black International, formerly somewhat restricted to the metropolitan areas, led by Germans, Bohemians and Scandinavians, began to feed on the depression throughout the country, extending into smaller cities, taking in the unemployed, including many native-born Americans in their ranks.

The Black International took the eight-hour-day plank of the Knights and demanded it as the first step in the revolution. In the middle of a strike of the McCormack Harvester workers for an eight-hour day, a bomb was tossed, the Haymarket riots broke, and the country coupled innocent and guilty together. Samuel Gompers was to say later that it not only killed a few policemen, it killed the effectiveness of the labor movement in America for some years.

It was all discouraging to a young man like John Mitchell who had been hoping for help from the outside. He was sixteen now. He went West. He would carve out his own fortune in a better, freer country. There were thousands of unemployed who made the same trek, homeless, disinherited men for the most part. Some eighteen years later the world was to know them and fear them under the name of the Industrial Workers of the World. In 1886-7 they had no organization quite so adapted to their itinerant casual type of labor.

This getting out into the world, away from the hometown boys who had sneered at him for hanging up the wash, who had called him a sissy because he would not drink, who had laughed at him for a credulous fool, with whom he had not been able to keep up in school, was probably the wisest thing he could have done if he had been planning his career with the wisdom of an adult labor leader. Hardly any of the wayfaring men of the day had any money. Mitchell did not. Traveling was a matter of riding the rods or the bumpers, or walking. All belongings went into one small bag to be jealously guarded day and night. Some of the men turned tramps long before Colorado was reached.

Colorado at that time was no El Dorado. There too the mining camps were overcrowded, men competed against each other for the work. A sixteen-year-old boy was only a weak competitor. He finally did secure a job, and then found that he had to relearn his whole trade as the character of mining was different from the one he had grown up in at home. For a while he had luck in that he lived with the family of old Dan McLaughlin, from his home town of Braidwood. The old man's plans for a collective agreement of the operators of the four big coal states with the miners had fallen through and he had come out to be superintendent of a newly opened mine at Starkville. From him Mitchell could learn what undying energy some men bring to the hope and work of their lives.

For two or three years he worked in various camps in Colorado, New Mexico and Wyoming. For him there was no gold in those hills. He turned eastward. Some rumor of new mines opening at La Salle and Spring Valley in Illinois had come through. The life was the same all over. Perhaps it would be a little better now in Illinois. He reached home in the financial condition that he left it—penniless. But now he was a man who had been around a bit, a different, harder person than the boy who had left home, seeking change and fortune. He had reconciled himself to the miner's lot in Illinois as his inevitable destiny.

That acceptance bound him up with the fate of the Illinois miners, who were inside of ten years to have the strongest district of the most powerful labor union in the country.

II

LIEUTENANT OF REVOLT

IN the next ten years there was not enough food in Illinois to put flesh on a miner's bones. John Mitchell's childhood had kept him thin. He was a man who could and later on did take on weight. During these years he was slight, wiry, one hundred and twenty pounds of weatherbeaten muscle, giving no promise that the time would come when success would add another sixty pounds to his bulk. In these years he gave the impression of an ascetic. His hair lay down black over his high forehead. Keen, quiet brown eyes in a swarthy face occasionally misled people into taking him for an Italian. Shortly after his return, when he swept on with the sweeping forces of revolt, and wore the ministerial black which characterized all men of the times who considered themselves in public life, there were many who mistook him for a crusading Catholic priest. What he lacked in impressive height was made up somewhat by the poise he brought back with him from three years of hard buffeting in parts of the country which the miners of Illinois had never and would never see.

The miners' leaders were undergoing a process of unnatural selection. Everyone of them who had hopes for his fellows and for himself crashed full steam into one of two wrecks. The coal companies were not interested in encouraging or fostering any private initiative other than their own. When a man showed his head above his fellows he was dismissed, forced to go on the road, hunting for a job where all southern Europe had poured in on the same quest. This was the fate that met William D. Ryan, who was thrown out of Braidwood, Mitchell's home town, and had found another precarious hold on life railroading. Later he was to become Mitchell's most powerful backer. Those leaders who were not eliminated by the employers were more than likely to go under in the constant series of failures which marked this decade of defeat. The loss of strikes had to be blamed on someone, and man after man fell a victim to his fellows before he fell a victim to the discrimination exercised by the employers. The men were too desperate to be loyal. They wanted victory too much to forgive. This process had drained off some of the best men.

When Mitchell came back to Illinois in 1888, he found work for a short time at Spring Valley on the Illinois River, fifty miles southwest of Braidwood. It was one of the many towns flooded by Italians, Poles, Austrians and Hungarians. For a short while the dazzling sun of prosperity had touched Northern Illinois. So many men had flocked to its warmth that the operators felt free to set them to working against each other, bidding against each other for jobs, driving the wage bill down to the place where competition with the low-cost mines of Southern Illinois would remain possible. The threat of retaliation on the part of the miners was met, in Spring Valley, by a lockout.

It was a bitter welcome home for the nineteen-year-old

traveler. The miners in the other towns were asked for relief. They were on strike themselves. In July the companies started evicting the men and their families from the company houses and property. The correspondent of the New York *World* reported that men were "dying to escape slavery." He said: "Twelve hundred heads of families of Northern Illinois had not had a stroke of work since May; seven out of every ten families are sick, seriously so; malarial fevers, diphtheria, cholera morbus, ague and pneumonia form the bulk of the ailments. . . . Salt pork, potatoes and corn meal, with a little tea and coffee, have been their sole means of subsistence throughout the lockout."

Men looked for help to their unions and looked in vain. The two rival organizations were carrying on a struggle for power, too intense to allow of any compromise with the rival. The only men who profited by their existence at that moment were the operators. In Spring Valley they forced a reduction of twenty per cent on their men, men who had during the months of employment previously earned between $45 and $60 —not including deductions for powder and tool dressing. Their leaders were blacklisted and all men who entered the mines were forced to sign yearly contracts of work which Henry D. Lloyd of Chicago—already noted then as an humanitarian—characterized as "slavery—slavery in yearly installments."

For the men of the mines the tradition of defeat was growing stronger. The loss of the struggle in Northern Illinois after almost a whole year of intense resistance came after the anthracite strike of '87 had been lost and after the miners in the Hocking Valley of Ohio had been beaten in '84. Even the men in the most isolated mine towns had learned one lesson: the time was overripe for a united front.

Pressure from below brought the two organizations—National District No. 135 of the Knights of Labor, and the other, now known as the National Progressive Union—together at Columbus on January 22, 1890. Ten long years of rivalry had created hatreds and ambitions which could be suspended for one short moment of enthusiasm but not stamped out until many men had gone to their graves. The rank and file, who did not believe it possible that men should quarrel among each other forever for power when there was no bread in the house, cheered themselves hoarse and cried with hope as the leaders of the two factions sunk, for the moment, their differences and the United Mine Workers of America came into existence.

The name was a proud boast. The mine workers were not united and the union did not cover America. It gave men's hopes a spur. Many of the younger men, Mitchell among them, joined the local branch of the newly organized union. They soon found that if they were to have better wages and conditions they would have to lead the fight themselves—the national organization could give them no help. They were the ones who had to take the brunt and risk of leadership.

In 1891 Spring Valley was on strike again, a strike which lasted ten months, the second one inside of three years. It was not a strike fought through with great hope, with any exaltation coming from the belief in victory. Neighboring mine camps which were solicited to make the strike solid and effective throughout the district made promises and then broke away. There were jealousies, bickerings, accusations of treachery. Some attempt was made to blame the foreigners for this repeated failure to make the unity as real as it was nominal. It was an explanation many accepted, but in the state and district conventions the miners represented in the wrangling bore the names of Murray and Crawford, Donnelley and Gallagher.

These were days which made a man want to be anywhere else than in the mines at the same time that they took away his ability to move out of the mines. It was a day when men looked for panaceas, hunting them in Darwin, Spencer, and Edward Bellamy's *Looking Backward*, a few years later rising to the challenge and promises of free silver. Mitchell was one of those who read *Looking Backward*, given to him by Father Power, one of the several priests whose lives were identified with the struggles of the miners. With some encouragement from a local lawyer, he began to read law. That was the way out which led to the legislature. Together with other men of his own age, in the union hall or in the friendly barroom of James Hicks' saloon, they discussed "national issues," everything and anything which seemed to give promise of bringing prosperity into the poverty-stricken world they knew. A lone survivor of that group still remembers the "wonderful big words" of the young mine leader. He never went on with the law. At best it gave him some self-confidence with his own people and helped him in 1896 when he was to go as legislative representative of the miners to Springfield.

The ten months of strike in 1891 ended in defeat again, and again John Mitchell trekked out West to try his fortunes. New Mexico proved no kinder than Colorado and in 1892 Mitchell was back in Spring Valley. He gave definite signs of settling there now. In 1892 he married Catherine O'Rourke, daughter of what one writer, with unconscious irony, called "a prosperous miner." She, too, knew the hardships of mine life. She, too, was motherless and she accepted the lot of miner's wife philosophically as did most miners' daughters. He was twenty-two then and had spent most of his life boarding around. She was two years older, she was stable and calm and marriage to her signified a certainty he had yearned for. They moved into the double-family house which his father-in-law owned, occupying it jointly with another family. Within the year their first child, a son, was born. He had a stalwart faith in the future.

There was little on the horizon to justify it. That year the state union only received $187 to do all its work. The anti-truck law passed by the State legislature, in which the miners had placed so much faith—abolishing the system of payment in scrip and compulsion to trade at the company "pluck-me" stores—was declared unconstitutional. The union, the courts, the State, politicians and the business interests of the country were not helping the miners any. They turned from one to another, looking for relief. The panic of 1893 swept all but five hundred men off the State union roll. Everything seemed to split them against one another, religion, national origin, the black record of dissension. The only thing they had in common was hope, and that was food for no one.

The country, many historians agree, was stumbling towards revolution. It was not the revolutionists who were making it. It was men who had been beaten so often in their peaceful attempts to gain the bare bones of existence who were becoming convinced that there was nothing left to them but violence on a large scale.

The miners were by no means alone in desperation. In the year Mitchell married, the best organized union in the country, the steel workers, went down under Carnegie and Frick in Homestead. Pinkertons shooting workers, Berkman shooting Frick, came into the scene of possibilities. In later years many miners would forbid their children the Carnegie libraries. The switchmen in Buffalo, coal miners in Tennessee, metal miners in the Cœur D'Alene, all told of the power of the great industrialists and the "lank, stark, naked failure" of the workingmen to protect themselves.

In that year the People's (Populist) Party, breathing the spirit of radicalism, polled over a million votes, captured twenty-two presidential electors, and sent a powerful delegation to Congress. Mitchell cast his vote for the Populists.

In 1894, a little before the Pullman strike began its course, a little after the American Federation of Labor had endorsed nationalization of the mines, railroads and telegraph systems, the President of the United Mine Workers, John McBride, stood before the fifth annual convention of the Union, with the simple proposition: "There is a limit to human endurance and you have reached that limit." He said it to a union which had only thirteen thousand paid up members out of the two hundred and fifty thousand men working in the mines of the country. The first call of the national union for a strike, to begin on April 21, 1894, was obeyed by one hundred and twenty-five thousand men. It met the welcome men reserve for audacity.

People were beginning to shake in their boots. The miners burned bridges over which the coal trains came up from West Virginia to break the strike. People remembered the Molly Maguires. President Cleveland called out the troops at Chicago to put down the "rioters" in the Pullman strike. Coxey's army was in the making. Two years were still to intervene before Bryan could declare to the workers of the country that their relief lay in political action, that he knew one way in which labor should not be crucified.

In Spring Valley itself, where Mitchell had been elected to the Board of Education, one of his fellow officials, the county charity officer named Lovejoy, advised the miners in August of 1894 to accept the reduction in wages. If they could not make a living at the reduced wages, he said, the country would make up the deficit. This was a flick on the raw of pride, all that the miners had left. Their retort to that came in the *Mine Workers' Journal:*

> "If after toiling and moiling, sweating and fuming, digging, backing, shoveling, sawing, smothering, gasping, scrapping, in water, mud, bad air, foul stenches, dangerous caves, treacherous rocks, pushing, heaving, twisting, hammering, lifting, prying, boring, wedging, sometimes in spaces not sufficient for full-grown rats, sometimes where the earth is open like a vast toothed dome whose teeth are hanging in the shape of jagged stones and ready at any minute to crash down and kill—this man Lovejoy, a public officer, tells those men that if they, after doing all this from ten to thirteen hours a day, and after having exhausted every nerve and muscle and vein in their body, have not earned enough to pay for the supper, the county will help them—that is, they will sup poorhouse fare after toiling like galley slaves. . . ."

Such pride did not keep men who had often been beaten from a complete rout. They went back to work "through with the union." In Spring Valley Mitchell and his wife saw the president of the local union sent "on the tramp." He was one of the many who had worked hard during the strike and had met the fate of all those who did anything for the miners. It was something for a young married man with a newborn baby to think about.

While Coxey's army was starting from the mining camp of Massillon, Ohio, in the most prominent of a long series of attempts which the disinherited of the great industries have made to appeal to the Great White Father at Washington, Mitchell was learning just how complete a defeat can be. He had struck with the others; he had stood by the others; he had thought that defeat could be no more than hard work for a pittance. When it was all over, he did not even get his job back. The new superintendent in his mine was an Englishman. He immediately discharged all the Irish. Mitchell, who was not quite Irish, but who had married into a Catholic family, was one of those to go. He found work in a near-by camp. Here he was found by an old acquaintance from Braidwood, the William D. Ryan who was later to play Mark Hanna to his McKinley. Ryan, who had been out of mining for three years and had worked at railroading, was traveling around as secretary of the sub-district of Northern Illinois, the only one then in existence in the state. Mitchell's senior by ten years, he picked the twenty-four-year-old miner as a man with qualities of leadership. At a time when so many of the miners' leaders had been pushed out by their employers or by their fellows, he chose Mitchell as one of a staff of men still willing to attempt to organize the chaos.

In Spring Valley that year, on Independence Day, three hundred miners made an offer to the world to go into voluntary slavery in return for reasonably comfortable homes, plenty of fuel and food and serviceable clothing. According to a Cleveland paper:

> "They represent the best elements among the miners and are willing to serve thus without a cent of wages. They say they will sign an ironclad contract . . . During the past three years they have suffered so . . . that rather than see their families suffer any more they will become serfs."

Illinois was not alone. Miners' children in the western fields of Pennsylvania were driving dogs away from garbage so that they might search through it for food. A legislative commission of that state reported that miners there "lived in many cases worse than beasts . . . herded together like cattle and in many cases wallowing in their own filth." Ryan stated that wages in Illinois averaged $12 a month, and that the miners were in worse shape than they had ever been in since coal was found in Illinois.

There was need for men to work hard and fast. Just what could be done that had not already been tried nobody knew. To be inactive was the only form of surrender that was recognized for itself. Mitchell became secretary-treasurer of the sub-district organization, which meant being the main financial official of all Illinois. At one of the first joint conferences of operators and miners of the sub-district, he heard the miners' stories of starvation, unemployment and injustice matched with the operators' stories of harsh competition and of "wild-cat" operators who would have nothing to do with the union, over-expansion, discrimination of railroads against union fields. It was the first of many, many times he was to hear both sides complain of a system over which they had no control.

Bryan, who had been endorsed by the Populists in '96, went down to defeat before Mark Hanna's candidate, McKinley, who had gained some repute amongst the strong Ohio miners' union for defending strikers before the courts. Mitchell went to Springfield as legislative representative of the Illinois miners. Here he helped secure—according to a correspondent of the *Mine Workers' Journal*—an amendment to the anti-trust law which gave operators and miners the right to go into convention and fix a scale of prices, something some of them had not dared to do before. The Legislature also passed a weighing law, a child labor law, and an act establishing an apprenticeship in the mines of two years. It was a position which made him somewhat known throughout the state. Some of his friends spoke of sending him to the legislature. A career in politics had lured other miners. John Mitchell had neither the talent nor love for a political career.

Ryan needed him in the state union. The organization was still weak. Good men, men willing to serve without salary, to trudge over the state, to spend night after night at meetings, to face adversity and opposition, were scarce, and Mitchell's personality, his willingness to serve, made for a rapid rise in the State organization. In 1897 Mitchell became a member of the State Executive Board, after having been defeated for the vice-presidency by an older, more experienced man. Ryan was elected State secretary-treasurer. It was an effective working combination.

John Mitchell was twenty-seven, and board member in the organization of one of the weaker states of the miners' union, when the national organization, on July 4, 1897, declared what was to be its first victorious strike. This time one hundred and fifty thousand miners—one-half the total in the bituminous industry—joined "the spontaneous uprising of an enslaved people." Less than ten thousand of them were members of the United Mine Workers.

The national organization had neither a tradition of victory nor leaders to whom it gave unlimited confidence and trust. In 1890, the A. F. of L. had selected the carpenters' union as the first to make a nation-wide struggle for the eight-hour day, and the carpenters had been victorious. In 1891, the miners' union was to be the second organization to make the attempt, but it felt itself too weak, with its twenty thousand membership, to accept the responsibility. Three years later, in 1894, with a larger membership, the union had called a suspension which had ended disastrously. It was powerful enough in the councils of American labor to have its president, John McBride, defeat Gompers for the presidency of the American Federation of Labor some months later. McBride's successor, Phil Penna of Indiana, had struggled on with a beaten union for a few years, and had resigned finally to take the labor commissionership for the Indiana coal operators.

When Michael Ratchford took office, in 1897, the union was so poverty-stricken that there was not enough money in the treasury to call a meeting of the national executive board and the district presidents. Demands for increases, brave enough in print, had been formulated at the convention, but hardly had the delegates reached home before "coal took a tumble in the market" and wage reductions were made in all coal fields. In the face of local unauthorized strikes, bound to end in failure, Ratchford attempted unsuccessfully to go into the matter of a national suspension by circular letter with his executive board. Finally, six months after his election, when the situation was desperate, he summoned the board and district presidents to Columbus, Ohio, the national headquarters, for a meeting on June 26, 1897. It was understood that each man was to pay his own expenses. It was at this meeting of a few men, representing an impoverished organization, that the decision to call the strike was made.

These men, with an inherent dramatic sense, called the strike for July 4th—Independence Day—just as their Illinois brothers some years back, on the same day, had offered to sell themselves in slavery. The decision to strike on July 4th was strategic also since no one expected the strike, and the operators were not prepared to fight. Previous strikes had always occurred on April 1st or May 1st, so that wage changes might be included in the reckonings of new prices by the operators. The call for the 1897 strike was not issued until a day or two before the appointed time. There was no time to build up reserves of coal or to import strike-breakers. Furthermore, the miners' leaders, not quite sure of their strength, had figured that the men would be idle that day, anyhow, and that the very meaning of Independence Day, plus the impetus of the day's idleness, would give a healthy start to the strike.

In Spring Valley, under the leadership of Mitchell and others, every miner responded to the call. Everywhere in Northern Illinois the men came out solid for the strike. Only four hundred and fifty of the poverty-stricken men had been members of the union before, but now, "the miners . . . procured bands of music and patriotic banners and marched from district to district to add strength to the strike feeling. . . ." Forgotten leaders came to the front again, and new leaders arose, volunteering their services, tramping from one mine camp to another, speaking, gathering funds. Chief amongst these leaders were the State officials—Ryan, Hunter and Mitchell.

Probably no one was more surprised at the astounding response to the call than the man who ventured the call, Michael D. Ratchford, national president of the almost bankrupt mine workers' union.

Altogether, the move was a bold and startling one. Yet there were some important districts which did not respond fully—West Virginia, Southern Illinois, Central Pennsylvania and the lesser districts of Maryland and Kentucky.

It was to the task of organizing Southern Illinois ("Little Egypt")—later to be made famous by the Pana Virden and Herrin Massacres—that John Mitchell was assigned, in addition to his work in the Northern District. Southern Illinois, with its "Little Egypt," with its Kentucky feuds, its influx of black labor, its traditions of American individualism, was a difficult territory to organize; Mitchell, and the men who helped him, succeeded in preventing any mass production of coal in the region. The defection of Southern Illinois in the strike would have led to its complete defeat.

By August, a month after the strike, it was reported in the *Mine Workers' Journal* that forty thousand Illinois miners were out, including the aristocrats of the industry—the coal-hoisting engineers, who belonged to another organization. All through the hot summer months, the young miner from Spring Valley tramped through the woods or along the railroad tracks, without money in his pocket for train fare or else without the price to hire a hall, reaching out-of-the-way mining camps with the strike call. Meetings were held in the woods when no hall was available; where a hall was available, there was always that crucial moment when money had to be raised from the poverty-stricken miners, not only for their own needs, but in order to pay the rental.

Without a doubt, the strike was one of the greatest struggles in American labor history up to that time. It enlisted the sympathy of the American Federation of Labor and of the still existing but much weakened Knights of Labor, of Eugene V. Debs of Pullman strike fame, of Mother Jones, of W. D. Mahon of the Street Car Men, of James O'Connell of the Machinists, and many others.

When West Virginia began, in 1897, its long-to-be-continued record as the state that used injunctions to prevent unionization, the most prominent labor officials of the country poured into the state in protest. West Virginia was indeed, then as now, a problem and test for the Union. There was little or no trace of unionism in the state. Organizers sent there found themselves confronted not only by the isolation of the mountainous camps, by the barriers erected by company-owned roads, but by privately hired police and by injunctions forbidding them to trespass on company property—and the highways and streets of many mining districts were company property. Although Mother Jones and other organizers were arrested and flung into prison, the strike gained some adherents in West Virginia.

All in all, Ratchford and the other leaders of the organization had cause for content. Public sympathy in the cause and leadership had been aroused to such an extent that when strike leaders went about collecting relief funds in the cities and farming districts, Governor Tanner of Illinois "publicly declared that the conduct of the strikers was praiseworthy; that they had learned the secret power of all great undertakings . . . self-control. . . ."

To many of the labor leaders and sympathizers with labor, the strike afforded a great forum for the discussion of ways and means of advancing the position of the worker. Meetings of all sympathizers were called by the American Federation of Labor to devise means of aiding the starving strikers. The first of these was held in July in Pittsburgh; the second, as a protest, at Wheeling, West Virginia, itself, at which latter meeting, according to the *Journal*, Samuel Gompers described the strike in his characteristically dramatic way: "A wail of anguish, mingled in desperation, arises from the bowels of the earth, and the miners' cry for relief, for some degree of justice, touches a responsive chord in the hearts and consciences of the whole people."

Mitchell was not able to attend these two meetings, but the third of them was held in St. Louis—only a hundred miles or so from Springfield. Mitchell and Ryan had enough between them to pay the fare, and they were willing to hazard the chance of food and lodging. . . . At this conference they heard trade unionists, members of the Knights of Labor, Social Democrats, Free Silverites, each with his own philosophy, expounding not only means of aiding the strikers but of bringing about the millennium. After several days of listening to speech-making, and several nights in twenty-five-cent "flop" houses, John Mitchell and the miners' delegates returned to the field to carry on the fight.

This was the period later to be called the Gilded Age. The victories won by the new industrialists on the railroads, in the steel mills and coal fields, were bringing money to a young aristocracy, which had not yet learned that to spend discreetly is to spend well. The newspapers carried stories of favorite steeds being fed champagne and roses; of a pet monkey which had its own valet and carriage; of Roman dinners at which guests found cigars wrapped in hundred-dollar bills; pies out of which chorus girls hopped; of opera glasses costing $75,000—these were the items which the impoverished miners read. The grand climax of the period was in the Bradley-Martin ball of 1897, held in the Waldorf-Astoria, when that hotel was converted into a replica of Versailles and Belmont wore a suit of gold armor inlaid with gold, valued at $10,000, and other guests, expended similar fortunes which might have fed whole mining communities.

The strike was an unforgettable lesson in leadership for the future president of the miners. Public opinion came to the aid of the soft-coal miners now, when it had failed them and the railroad men in 1894. To some extent this change in the temper of the country had been brought about by the agitations of the Populists and of the Boy Orator of the Platte. The continual emphasis on law and order by the miners' leadership had impressed those who recalled the Haymarket, Homestead and Pullman riots and the bridge burning of '94. Not even the accusations of political motives behind the coal struggle—accusations which were thereafter to be made in every coal strike—alienated public sympathy. Perhaps it was the oncoming of a change for the better in the economic conditions of the country as a whole.

By August, when over five thousand West Virginia miners had struck, the operators who in July had "nothing to arbitrate" began to make overtures for peace. The settlement offered was below the scale decided upon in the strike call, and was not on a national basis. The miners, now aware of their strength, refused the offer. Early in September a conference of the district presidents and the Executive Board was arranged in Pittsburgh. A more favorable offer of the operators was submitted, though it was still a compromise. Its chief attraction was a provision for a joint interstate conference of operators and miners which was to meet in Chicago to decide on a new scale in January of 1898. After much discussion and opposition, the offer was accepted and most of the men went back to work. It was the first national victory the miners of America had ever known.

Curiously enough, the settlement was not sufficient for the miners of Illinois. They had been recalcitrant members before the strike. Now they were thoroughly aroused and refused to return on the scale accepted by the National Board and the district officers. They had built up a large and militant organization. The settlement was spoken of by the "hotheads" as a betrayal. Mitchell, who was to prove later more completely that he was by temperament a conciliator, defended the national officers and went out campaigning for unity. He knew the need and value of discipline. As late as October, however, he was still out in the field trying to get local settlements. When his efforts failed, a joint meeting of the operators and miners of the northern section of the state was called. At this conference, Mitchell's was one of the few voices raised for peace. In the absence of the district president, he acted as chairman of the joint session. The miners of Illinois were still amazed to see the man who had had so large a share in organizing the difficult sections of the state pleading for what seemed to them an unreasonable compromise.

In November, two months after the general settlement, the Illinois miners and operators arrived at a scale higher than the Columbus settlement, ten per cent in advance of the scale of the previous May. Peace did not come, however, until the state president, James Carson, was suspended on sensational charges of "selling out." His office as president was taken by John Hunter, a former preacher. His coveted place on the now powerful National Executive Board was filled by Mitchell's friend, W. D. Ryan. For the moment Mitchell remained only a member of the State executive board; a few months later he went as delegate to Illinois, to the National Convention. It was a not unimportant achievement for a young lieutenant of the revolt with only two years of official service and twenty-seven years of life.

III

THE FLUSH OF VICTORY

THE long monotony of local strikes and local defeats was broken at last. Old Dan McLaughlin's ideal of treating the labor problems of the national industry on a national basis was nearing realization. Hope excited men like drink. Some men had starved, other men had gone bankrupt because of the long failure of the industry to accept the proposition that stabilization is better than chaos, peace better than war, compromise better than unrestricted private initiative.

To the miners it became a question they had never met before—to put flesh on the bones, facts on the promises, of their triumph of 1897. A few things they had learned: courageous local attempts to coerce employers into giving the wage the miners wanted were expensive, futile gestures. Other operators and other miners would simply take the advantages of such local strikes with glad and selfish eagerness. Another thing: the four states in the central competitive area—Illinois, Indiana, Ohio and Western Pennsylvania—produced only one-third of the coal of the country. Any defection in the unity they were still striving for in that area, any extension of the notably non-union production of West Virginia or Kentucky, would shake the whole plan for stability in the industry. They had also learned that the steady importation of workers from Southern Europe, the policy of unrestricted development of every coal seam discovered meant that they could only work two-thirds of the time, that the industry was fifty per cent overdeveloped. They had to have a shorter working day to spread the work around. They had also learned, from bitter repetition, the difference between reality and promise. No matter what scale of tonnage rates the operators promised, there were always ways and means for them to nullify it. The most common technique was a process of enlarging the holes in the screens over which the coal was run to eliminate the dust and small unmarketable chips. The men knew that unless they insisted on getting paid for the "run of mine" coal (all they produced) or in interfering in management to the extent of fixing the size of those screens, they would always be at the mercy of their employers.

Now they were meeting with operators who had been beaten without being completely convinced. In a world where the word of the operators had long been law, where their mine guards had been the police and the final authority, where union leaders had been dismissed as a matter of course, those leaders who had survived were attempting to protect their people and themselves against any renewal of the continuous widely scattered sniping. Every point they gained was so much more bread and butter, so much more guarantee that they would no longer have to live in fear of their jobs, so much more opportunity to stand up and talk out like free men, fighting for their stake in the country, their right to share with others the covenants of life, liberty and the pursuit of happiness.

In convention assembled at Columbus, Ohio, on January 11, 1898, the delegates from the mine towns knew all these things. They knew too that fortune was favoring them at last. Prices for coal were rising. A few months later the Spanish-American War was to help an industrial boom. In the State House some blocks away Mark Hanna, who had made McKinley, was fighting for his own place in the sun, a United States senatorship. The leaders of the Ohio miners were called upon for pressure on their assemblymen. The union began to see itself as a factor in national politics, capable of interesting both parties. The long hunger for success partook of every food.

The men knew what they wanted. There was no irreconcilable conflict. Without much discussion they turned the matter of formulating demands over to a scale committee and set themselves about a task which they felt to be more surely within their own field of experience and competence—that of choosing leaders who would really stand up and fight for them. Men they could understand and evaluate. The principles those men stood for, in working out the agreed program, the difficulties they encountered, were then and have since been objects of their lesser interest.

Many of the delegates wore broad-brimmed sombrero hats and long miners' coats. Their meeting was half-hidden in the clouds of smoke from the ever-present miner's pipe. For all the oppression they had undergone, they were an articulate, well-informed group, even though a few may have been somewhat the worse for liquor. Ten states were represented there, and though many nationalities were involved, the delegates elected were in the main of the older group—English, Welsh, Scotch and Irish—with only here and there a lone Slav or German or Italian. The spirit of the convention was elated and jubilant, despite an occasional note of doubt of the permanency of the victory or of fear that the spirit of unity might be lost again.

It was John Mitchell's first national miners' convention. The only other meeting of national scope he had ever attended had been the St. Louis conference, held the previous August to devise means of helping the strikers. He came to it as one of the younger men who had risen rapidly.

Before the delegates had time to congratulate themselves on their gains, the factional fights for office began. This time Illinois had one hundred and four delegates and one hundred and ten votes, in contrast with only one at the previous convention. Ohio, always the strongest union state, had one hundred and thirty-six delegates (one hundred and fifty votes), split, however, by rival ambitions. Indiana, Western Pennsylvania and the states bordering around the central competitive area made up the total of four hundred and thirty-one votes.

Each state promptly developed desires to be represented by national office. Ohio had the president, Michael Ratchford. Like Ryan and many of the leaders, he was some ten years Mitchell's senior. He brought a handsome, almost stagey presence, together with the first triumph of a national union, to the scene. He was one of a long series of good-looking men whom the miners had delighted to honor. A friend and partisan of Hanna and McKinley, he was shortly to leave the union presidency for an appointment to the United States Industrial Commission. After that he was to be State Commissioner of Labor, commissioner to the operators of Ohio for a short time, and to hold other minor political offices. Already there was talk of his retirement into the political field. In spite of that he was reëlected.

Ohio wanted more. A candidate for the vice-presidency was given a complimentary nomination, but withdrew. The anthracite field proposed Big John Fahy, tall, upstanding, with long, drooping mustaches, and a gift for the florid oratory of the period. He was called the handsomest man in the union. The anthracite fields were still very weak, however, and Fahy also withdrew, as did several Pennsylvania men. Ryan and Hunter of Illinois withdrew in favor of John Mitchell, who was then left alone in the field to oppose Patrick Dolan of the Western Pennsylvania field.

This was the first of a long series of clashes between the two. Hail-fellow-well-met, rough and ready, not bad to look upon, amusing the convention with his songs, Dolan brought with him some of the prestige that was given men who had served an apprenticeship in Scotch unionism. He had proved that he knew the way to power. District president, member of the National Executive Board, he had all the prestige of political success. Against him Mitchell could only bring the appeal of his frank, youthful appearance, some popularity in the state of Illinois, and the ability of W. D. Ryan as a political strategist. At the beginning of the convention, Dolan's standing was recognized by a place on the most important committee, the scale committee. Mitchell's was recognized by a place on the committee second in importance, the resolutions committee.

Men were looking out for themselves and their friends for some years ahead. The possibility that Ratchford would resign made the vice-presidency important. The men in Ohio who had been fighting Ratchford wanted his job. Was the vice-presidency the route to that job? Ohio could not have the two chief offices at once. Better to take the promise Ryan offered that in return for giving Illinois the vice-presidency now, one of the Ohio men—Tom Lewis most probably—would get the Illinois support for the presidency when it became open. Dolan, counting on Ratchford's support, did not see all these long-term plannings. He knew he had Indiana and Pennsylvania. He let the vote be taken, and found that Mitchell was only one short of the necessary majority, receiving two hundred and fifteen to his own one hundred and ninety-five. The scattering votes switched to the leader. Dolan picked up only one but Mitchell picked up thirteen and was elected to the vice-presidency of the United Mine Workers.

For one brief moment the bad blood caused by Dolan's defeat, the hot accusations of double-crossing, were forgotten. In a burst of jovial song, with Pat Dolan leading a somewhat tipsy circle in the strains of "The English, the Irish and the Scotch," the convention adjourned. The chosen delegates took the special train to Chicago to meet the operators and to prove to their people and themselves that the hope centered on them could be justified.

At Chicago there came into being an arrangement which was

to be the main stabilizing factor in the industry for an almost unbroken period of thirty years. It was the first national agreement which any important industry in the country had made with its men. It represented neither dictated peace nor bankrupt surrender. The four states competing most directly with each other agreed that the one item on which they would not compete was labor cost. The operators protected themselves against financially weak or unscrupulous competitors. The miners protected themselves against their own people in the labor market. The first meeting of what was later to be the "annual joint conference of coal miners and operators of Illinois, Indiana, Ohio and Western Pennsylvania" did not move either swiftly or surely toward the function it was to fulfill. Too much material for controversy had accumulated. At moments it seemed as though the sectional competition between the operators of the four states would be too strong a barrier to overcome. The operators had at best a loose temporary organization without compelling discipline or responsibility. They had to gamble on the power of the union to enforce the agreement.

There were times when the miners too seemed linked to the operators of their district, their work-givers, more than to their fellow-miners in competing states. Conflict between Dolan and Ratchford was discussed in the papers. For days the scale committeemen met in secret and were able only to disagree. Time after time Ratchford arose to speak—a living reminder to both sides that the miners could be led to a victorious strike, a red flag to the defeated candidates for office. It was a training ground, full of pitfalls for the young vice-president of a national union. Behind him was only one element of comforting strength—his Illinois miners were solidly organized; no matter what happened to the joint conference they at least would get the eight-hour day and payment on a run-of-mine basis. At public meetings he was able to stand up and advise coolness and patience.

The miners' leaders did not allow the victory to slip away from them. After eleven days of wrangling, they were able to take home to their expectant people a further increase in wages, the pledge of the operators to collect the dues for the union from the wage bills (the check-off) for everywhere except Western Pennsylvania, where the union was weakest. The eight-hour day was won. Illinois coal was to be paid for on a run-of-mine basis. In the other states the size of the screen was limited.

It was great news to the people at home. For once they had the sensation of being able to tell the operators what they wanted and of making it stick. When Mitchell came to Spring Valley, to his wife and children again, he was met by a parade, speeches and praise. The local boy had not only made good; he had made the union good.

At last he was out of the mines. He could put away his pick, cap and lamp, overalls and boots, and wear the broad-brimmed sombrero and traditional long coat of the leader. The days when he had to crouch on his haunches or lie flat on his belly ten hours a day in a three-foot vein were over. From the work he knew how to do well enough to support his family and himself, he had to move into a new and strange world and learn work which was to support forty thousand other miners' families. It was a traveling job and meant the beginning of a long separation from wife and family, from the brief interval of stability he had found for himself in a wandering, lonely life.

Mrs. Mitchell accepted the separation as she had accepted the hardships of her life. She was absorbed in her family—there were now three children—and the uncertainties of political office hardly warranted giving up the established life in Spring Valley. She had been proud of her young husband as a local mine leader. She would have been content had he stayed as State leader. She was proud now that he had won national honors. His salary of nine hundred dollars held the promise that his children would have a less precarious life than he had had himself.

There was no time to rest at home on his laurels. West Virginia, with cheap non-union labor, was taking business away from the Ohio operators. There was a threat of a break in Southern Illinois. The union, although grown from ten thousand to forty thousand men inside of two years, was still contending with two hundred and thirty-one thousand unorganized men in the bituminous industry. Discipline had to be built up. Men had to be convinced that the union had a power which it had never had before, a power sufficiently sustained to make unnecessary the greedy seizure of what they could get in years of prosperity.

To the young vice-president was apportioned the task of organizing West Virginia. Although five thousand of the fifteen thousand West Virginia miners had answered the strike call in '97, there had been no victory in that state.

Wholesale arrests, injunctions, evictions and deportations had broken the men. Mitchell's stay was not lengthy; almost immediately he ran into the private police system of the state which had won for it the title of being the stronghold of feudalism in America. An organizer of the national office who had gone down to West Virginia with Mitchell tells of their having been pursued in one company-owned town by armed guards who shot at them, and of their forced swim in an icy mountain stream in order to escape death.

Mitchell could hardly be said to have been successful in his mission. At the end of 1899 the membership of West Virginia was only three hundred and seventy-five. The national office, as an alternative, resorted to the expedient of sending out a circular to organized labor and its friends everywhere asking them to boycott West Virginia coal. John Mitchell was sent off to other duties. It was defeat for him, his first.

The organizer mentioned above tells another story—for which no confirmation can be found. According to this, Mitchell was sufficiently discouraged after the West Virginia experiment to go to Ratchford with his resignation. Ratchford induced him to stay by holding out the possibility of the presidency in the event that he, Ratchford, should accept a post on the newly created Industrial Commission.

There was danger nearer home. In Southern Illinois, where the miners were less strongly organized than in the northern part of the state, several individual operators were breaking the terms of the national agreement, and attempting to run their mines on a non-union basis. Mitchell was sent into the situation. He spoke at length of the sacredness of the contract, but even the State Board of Arbitration suggested a compromise below the scale in that contract. The operators not only rejected that, they began importing carloads of negro laborers from Alabama and Tennessee. Southern Illinois, which had been settled to a large extent by men from Kentucky, men accustomed to white supremacy and feudal practice, saw the prospects of riots and racial wars.

Mitchell and Ryan sent hurried wires to Governor Tanner, who had shown himself favorably inclined to the miners' cause in the strike of '97, to intervene. Tanner sent David Ross, ex-miner of Braidwood and friend of Ryan and Mitchell, and then Chairman of the State Industrial Commission, to investigate the situation. The Illinois leaders sent word to the organizers in Alabama, but trainloads of negro miners, unsuspecting strike breakers, kept pouring in. These trains were met by the strikers and their leaders, and in many cases the negroes were persuaded that it was not at all for their welfare to remain in Little Egypt. In spite of all advice as to moderation, feeling amongst the strikers ran very high.

Governor Tanner stopped further importations of negroes on the ground that many of them were probably criminals who had learned their trade under a convict system, and sent out the State militia with orders not to permit their disembarkation, the first time in the history of the State and of the nation that the military power of the law during an industrial contest has been exercised in defense of the rights of American labor.

It was a moment when the responsible leaders—with the eyes of the whole state on them, with the inner knowledge that if the Pana Virden field were lost, on top of the failure in West Virginia, the whole interstate agreement, and with it the union, might drown—needed to work fast and shrewdly. Mitchell's friend, W. D. Ryan, himself tells that, tired of public negotiations, he made a secret agreement with the operators to accept a lower scale on condition that all strike breakers be immediately sent out of the territory. He kept this from Mitchell. No sooner were the negroes on the south-bound train than negotiations for peace began again. Ryan repudiated the secret agreement and the operators, outwitted, finally gave in.

Later, in personal conversation, Ryan confessed and asked Mitchell what he would have done had he known of Ryan's part in the secret agreement, even though Ryan had done it in shrewd anticipation of the ultimate outcome. Without a moment's hesitation Mitchell, according to Ryan, turned on the man who had been his friend and strategist for the last five years, and said: "I would have denounced you and demanded your expulsion."

The feeling engendered continued with the lockout all through the summer. On September 2, 1898, sixty miners of Pana seized David J. and Levi S. Overholt, president and superintendent of one of the mines, and threatened to lynch them because of their vicious tactics in the fight against the miners. A Protestant minister tried to intervene and was badly injured. Someone sent word to Mitchell and Ross, who were having a conference at the time, that murder was very decidedly in the air. The story goes that Mitchell entered the angry mob and rescued the two men, not without personal risk. He and Ryan accompanied the rescued men to a near-by store, followed by the crowd who insisted on hanging them. Once the operators were safe, they promised to meet with a committee of the strikers, and Mitchell dissuaded the miners from further violence. Where he was not condemned for his folly, he was praised for his bravery. For a moment it wiped out in his own mind his failure in West Virginia.

The owners and men came together then, but peace throughout the district did not come until after the disastrous "Virden Massacre" in which mine guards employed by the companies fired on and killed eight strikers who were trying to prevent strike breakers from entering the mine. Ten mine guards were also killed. The "massacre" was the last straw; within a few days the struggle came to an end, the operators accepting the compromise offer made by the State Board of Arbitration. It was below the scale of the contract, but it was the best Mitchell thought he could get.

The Pana Virden struggle had more than a personal significance. It was not only a dramatic vindication of Mitchell; it meant the extension of the interstate joint contract to the Southern Illinois field—the most important section of this important district. The failure to make the operators live up to the contract had been a sore point in the joint conference in 1898. Had both this section and the West Virginia section remained unorganized, it is doubtful whether the interstate joint conference would have been continued. The work of years would have fallen to the ground, and Mitchell with it.

In the very midst of the Illinois difficulties, a meeting of the National Executive Board in Indianapolis, hastily summoned, was called for September. Ratchford had at last accepted an appointment to the United States Industrial Commission, which was investigating the causes of labor disturbances in the United States. There was much talk that a new election would be insisted on, instead of permitting the inexperienced young miner from Illinois to succeed to the presidency for the balance of the term. His brief stay in West Virginia had not been outstandingly successful; his conduct in Southern Illinois, however, seemed to assure as strong a following there as he had in Northern Illinois. Mitchell hastily left the scene of the massacre for Indianapolis.

IV

THE VICTORY OF YOUTH

OFFICE was one of the few means open to miners to secure some distinction for themselves in the community. Office within the union had the advantage over political office in dealing with things they already knew about, men and the mining industry. So much hard work had gone unrewarded for so long, so many leaders had gone away from the fields and had either been forgotten or never forgiven for failures for which they were only in part responsible, that those men who survived had an eagerness to be compensated in a quite different way.

Ratchford's resignation in the middle of his term seemed a fresh opportunity for all those who had taken part in the victory election of January. Rather than run the risk of repeating the political wrangles of January, it was decided by the Board, in the fall of 1898, that Mitchell should be Acting President until January of 1899.

Mitchell, with one failure and one somewhat compromised success to his credit, had now to organize his own victory, to overcome the objection men still raised to his youth, to meet Pat Dolan of Pittsburgh and Tom Lewis of Ohio on the battlefield of the first political principle, the ability to get elected.

There was need of unity in the organization. The agreement with the operators came up for decision again in January. The organization, still very weak in the outlying fields, needed to present its best front. Some things, however, had been accomplished and Mitchell, as Acting President, could take a share of the credit. Within the year Kansas, Iowa, Arkansas and what was then known as the Indian Territory (Oklahoma) had all made a beginning of organization. If it was pointed out that West Virginia still had its meager membership of two hundred and seventy-five, it could be answered on the other hand that Kentucky, now organized for the first time, had eighteen hundred members, that Tennessee and Alabama together claimed three thousand union men. There had been some increases of wages, some improvement of conditions. Part of the prosperity that had come to the coal fields could be attributed to the Spanish-American War, but the good feeling it had brought to the administration, of which Mitchell was head, could not be argued away.

Naturally and from childhood turning in on himself, owing the major part of his political success to Ryan's astuteness, a man of few intimates, Mitchell had a straight-out political fight for his life. Tom Lewis, who had supported him for the vice-presidency, had been promised by Ryan the support of Illinois for his own candidacy in 1899. To him Ratchford's resignation seemed premature. There were charges of an arrangement between Ratchford and the Illinois group to push Mitchell in for the presidency. On top of this, Dolan, who opposed Mitchell bitterly, announced his own candidacy. Dolan had the political advantage of having the convention meet in his home city, Pittsburgh. To that was added the argument that the district organization weakest in the central competitive area would be bolstered by having the president come from what was the largest coal-producing state in the country.

The fight opened bitterly. There were charges of extravagance, of using organization funds and machinery to secure delegates who would reëlect the men in office. The Lewis faction gained the right to examine the records of the locals electing delegates, a direct attack on allegedly padded membership rolls of Illinois. The Mitchell group in return made a fight over the credentials of John McBride, a Lewis partisan, who had formerly been president of the union and of the American Federation of Labor, but was now out of the mines.

Dolan withdrew shortly and the fight for control of the union narrowed down to two men, Mitchell and Tom Lewis. It was a fight that lasted as long as Mitchell was in the union and was not forgotten when Lewis finally became president himself.

Lewis, who was much the older man, was shrewd and capable. When Mitchell had been a youngster in the mines, T. L. Lewis and his brother, W. T. Lewis (Master Workman of District No. 135 of the Knights of Labor, and reputed to be the brains of the Lewis family) were well-known and influential in the miners' union. These brothers were two of eleven children born to a miner of Welsh ancestry, of whose family it was said by the Mitchell men that each was a Lewis campaigner in another section of the country. Lewis seemed to have the united district of Ohio behind him. A small man, with a large head and drooping mustache, he was an unimpressive speaker, a cold, unappealing, antagonistic leader and a bitter fighter. He was credited by some miners, however, with knowing more about the coal industry than any other man in the organization. He was well read, hard working, with plans for the education of the miners, for the elimination of the saloon and its replacement by lecture halls. He had a keen, logical mind, in fact, every quality of leadership except the most essential one—the ability to work with his fellows. His ambition made him suspicious of every man and from the beginning of his career he instituted a bitter factionalism, a spy system. He never forgot that he wanted to be president of the miners' organization.

Mitchell, on the other hand, was already in office and that in itself gave him a certain advantage. The record of achievement during his months in office, the voting power of Illinois —which now more than outnumbered Ohio and Pennsylvania together—the political strategy of Ryan, all these too were in Mitchell's favor. Finally, the personality of this quiet, self-possessed young man inspired a confidence in the delegates from the new districts which all the shrewd manipulation of the Ohio group failed to arouse.

Men were willing to give youth its chance. Shrewdness might lead to office, but the very inexperience of Mitchell in the jungle fighting of union politics led to trust. When the vote was about to be taken, Lewis withdrew and Mitchell polled five hundred and seventy-one votes, Dolan two, and a West Virginian who had remained in the fight forty. Mitchell's popularity was best shown in the election of delegates to the American Federation of Labor. He ran first and Dolan a bad fifth. This victory brought Mitchell the knowledge that from now on he could count on the opposition of Lewis and a portion of the Ohio group, of Dolan and the Pittsburgh group. He would either have to overcome it by sheer political mastery and dominance, or surrender to it by compromise and fusion, or trust to a splendid record of achievement to make his defeat impossible, or go under. In spite of his great majority, he had seen how close the roads to prominence and retirement may run. Back home a month later, he showed that the nation-wide acclaim which greeted the youngest executive a national union had ever elected, had not turned his head. He said:

> "I am proud to be back among my Illinois brothers who have stood by me when I was being criminated and recriminated (*sic*) at Pittsburgh until every charge had been refuted. . . . As your National President I shall never allow the trouble that I have passed in other states against me to make one iota of difference in my conduct, no matter how hard they may attempt to stigmatize my character. When I have finished my term of office, I shall return to Illinois and take up my pick among you. . . ."

The arrangement of the legal machinery of the organization had been of a character which allowed a political leader once in power to make, unchecked, almost endless moves to keep himself there. For all but two weeks in the year, during which the convention was held, the president was the supreme authority, checked only in case of dispute by the General Executive Board. He appointed all the national organizers, the traveling salesmen of unionism, and it was to him that they were responsible. As they went around the country, they could either build or undermine reputations and characters. The president also filled any vacancies occurring in the national office. He passed on the constitutionality of an act, when appeal came from a local or district organization. He had a financial check on all the activities of the national union. This structure gave him wartime powers all the year round. There was the check on the executive of an annual election—but that was all. In strike situations, he not only was accustomed to receive a fairly free hand from the district committee, his very presence in the situation made the public consider him as the final authority in the negotiations. In spite of the fact that the General Executive Board not only had advisory powers but could impeach the president, the rank and file were inclined to acclaim a president who was successful and hold him alone responsible in case of failure. The complicated responsibility of the office meant that a man had to be a splendid success or spend most of his time in successful attempts to secure his own reëlection.

In the joint conference immediately following his election, the miners had on their side a long unprecedented comparative prosperity. They demanded higher wages and better conditions and seemed to be willing to take their chances with another national strike. The operators retorted that the failure of the union to organize West Virginia and to force the operators of that state into the interstate joint conference meant that they had no protection. Especially bitter was the complaint of the hardship caused by the eight-hour day and in Illinois of the run-of-mine system of payment.

Whether the young union could weather another strike just now was a matter of judgment. Mitchell threw his weight with those who thought it better not to risk the attempt until the interstate conference was on a sounder basis. The joint conference of 1899 ended in renewing the previous agreement and established in the minds of its participants the ability of Mitchell to discipline his own group. The easily wooed approval of a public which always dreaded strikes was also his. The miners' new president made his first step as a conservative.

The middle of the year brought a sudden rise in the market. The groups among the miners which had accepted the renewal without an increase in wages with reluctance began to clamor for a revision. Their cause had not only been proved right, the conservative element in the union proved wrong, but they also saw an opportunity to assert their influence on the side of higher wages which might not come again in a long time. Mitchell had had an experience of importance which was somewhat similar. The effort of men to seize every momentary advantage to improve their agreements was within the experience of practically every local or district leader in the union. It was a perpetual struggle between militant opportunism and the less spectacular belief that contractual relations were in themselves important enough to warrant some sacrifice of immediate opportunities for the long term good. In Northern Illinois Mitchell had tried to make his own neighbors see that it was more important for them to keep the national agreement than to get a little extra money on account of being close to the Chicago market. He had failed then to persuade them; they had won a rate higher than that allowed in the settlement. Other men who had taken the same attitude as himself had been accused of "selling out." The accusation had not been made against him; he was willing to take his chances that his own people would still trust him in spite of the fact that their immediate interests were at variance with the long term interest of the union. He came out at once with the proposition that unless the contract with the operators were kept sacred, the union's reputation in the country would be lost. At the next convention Samuel Gompers rallied vigorously to his support. The old hand was laid on young shoulders. Not only for the miners but for unionism in America, sacredness of contract was a badge of respectability.

The real work was in the field. Headlines in the papers whenever he spoke, an election to a vice-presidency in the American Federation of Labor, acquaintanceship with some of its leaders, with some politicals, a personal friendship with an Illinois operator, Harry N. Taylor, and the criticism he received for it, were all parts of his own education. The fact that without a fighting disciplined group in every field the union would become a top-heavy shell was something he never allowed himself to forget.

He did not go out as a conqueror, or even as a political boss with control over the political destinies of all the men he met. It had been hammered into him by friends and enemies that a young man must first overcome the handicap of his youth. Only after that could his youth become an asset. As soon as he arrived in one of the miners' camps, he would go to headquarters, sit there and listen to what the local leaders, what the rank and file who might be around, had to say of the situation. He took time for every man and every idea.

Men who had put in twice his years of life into the service of the union, who were accustomed to knowing their own minds, were often disarmed by his willingness to listen and to take suggestions. He learned by listening; he had infinite patience and faith. Even when his decisions were adverse to them, some quiet friendliness about him convinced them they had not been overridden. His quiet self-confidence did not bar the impression that here was a man who wanted assistance in a serious job. He aroused a desire to protect him in men who were his subordinates. After that they were sometimes willing to believe that his wider contact in the union had given him a greater ability to judge well. His demeanor disarmed also those who were afraid that their new union would become what the old one had turned into, a vehicle for personal promotion. He seemed to be making no attempt at all to even build up enough political machinery to assure himself an honorable chance for reëlection. In spite of the fact that five of the ten members of the National Executive Board were not friends of his, he appointed Ohio and Pennsylvania men, friends of Tom Lewis and Dolan, as organizers. In many cases this meant that they would control the delegates from the regions where they were active.

At this time he had the good fortune to have a large part of the wearing work of office detail, which so often occupies an undue part of an official's life because of his very inexperience with it, taken from his shoulders. In 1899 Miss Elizabeth Catherine Morris became his secretary. Her contribution to his advancement was, of course, of a somewhat different type than that of the men mentioned before. To the young miner, whose expense books of the previous year had been kept in the faltering handwriting and unsystematic fashion of the self-taught workingman, her aid in the important routine of the executive office was invaluable. It was she who told him when he mispronounced the big words he occasionally liked to use, and helped to make grammatical and easy-flowing some of the long, winding sentences of his speeches of that day. Miss Morris had come from a decidedly middle-class atmosphere; she had studied with and been secretary to May Wright Sewall, an author, then well known as a leader of women's organizations and founder of the International Council of Women. The rough-and-tumble surroundings of the miners' headquarters was a totally new experience for the rather delicate, diminutive girl. But the personality of Mitchell and his enthusiasm for the miners' cause won her devotion to him, and through him, to the miners. Her loyalty to him, her ambition for him prompted her to keep a record of his activities so that he might be better able to summarize his work and to plan it. In the innumerable ways in which a capable secretary can help, she helped. It was she who remembered the names of the miners who came to see him or who talked to him at conventions, so that it might be easy for him to call each man by name and receive his everlasting gratitude. It was she who jealously watched the men she thought were seeking to discredit John Mitchell. Under his instructions, she saw to it that no letter from any miner or from anyone interested in the miners' organization went unanswered, and that no grimy pitman was slighted if he called at the miners' headquarters. In the enormous task before the miners' president in 1900 and afterward, he knew he could count always on her loyalty, her efficiency, her genuine assistance. She stayed with him for twenty years, until his death.

He could keep himself free for the intense work of making the union a life factor in every mining district in the country. It was no small ambition. In Michigan, Missouri, Iowa and Kansas, district agreements now carried provisions for an eight-hour day and for wage scales comparable to those in the central competitive field. In Kentucky, Alabama and Tennessee the miners held joint conferences with the employers and secured increases, ranging in some instances as high as thirty per cent. Even in West Virginia the organized miners in the northern section negotiated an increase which influenced wages in the non-union sections. In the anthracite fields of Pennsylvania—which had remained broken since the disastrous Knights of Labor strike in 1887—some eight thousand men found courage to join the union.

Not everywhere were gains made peacefully and as a logical consequence of the growing strength of the union. There were strikes even in the central competitive region, in the Hocking Valley of Ohio and the block-coal section of Indiana, when operators refused to live up to their agreements. The most outstanding struggle of the year was carried on in the newly organized southwestern region of Arkansas and what was then known as the Indian Territory. When attempts to negotiate peacefully failed, the miners in these two districts went on strike. The operators replied, in West Virginia fashion, by obtaining injunctions which prohibited the holding of strike meetings. The leaders of the miners, John P. Reese of Iowa and Albert Struple of Arkansas, were arrested when they attempted to address a meeting of strikers, to which strike breakers had been invited, and Reese, as an outsider, was sentenced to ninety days in jail. Mitchell had wired the American Federation of Labor and the heads of the various unions affiliated with it to protest to Secretary of the Interior Hitchcock. Hitchcock, in turn, wired Mitchell to "unite with the department in suppressing such contemplated violations of law" as holding strike meetings. The youthful president, who continually, in his speeches to the miners, emphasized the necessity for abiding by the law, replied to this wire in no uncertain terms that labor was entitled to the protection of the law:

> "It was our understanding that the constitution of the United States guaranteed all citizens the right of peaceful assemblage. . . . I cannot understand why mine workers have not as much right under the law to secure brass bands and march to a given point for the purpose of holding meetings, as any other class of citizens. However, I assure you that our organization will not countenance or tolerate the commission of any overt act or violation of law."

Wide publicity was given by the organization to the "high-handed usurpation of judicial authority" and protest resolutions were sent out by many unions against what Mitchell then called "those protectors of corporate interests." The militant strikers carried on the struggle for seventeen months, with the aid of strike benefits from national headquarters. The momentary loss, in August, 1900, of the Indian Territory field, was replaced a few months later with an advantageous agreement and a strong union.

Men were beginning to take some pride in belonging to an organization that had upset the tradition of defeat. *The Mine Workers' Journal* boasted:

> "The present membership is much in excess of the number of men under the command of General Grant at the memorable battle of the Wilderness, the greatest in the history of modern warfare. It is a great industrial army, marching in the vanguard of progressive trades unionism. . . . We can now say with pride that the U. M. W. of A. is the largest organization in the world. . . ."

Mitchell could feel some of this seeping up to him through the ranks, coming to him also from his peers, from a certain new deference in the press. He had in his two years in the presidency made no palpable errors. While in 1900 he still looked the picture of the "boy president," still slender, wide-eyed and humanly appealing, there was underneath the maturity of a man who had not only had the luck to rise rapidly but had already convinced himself of his ability to carry on the work, to meet the great responsibilities of a growing organization, and a growing situation, in full. The difference showed itself even in his language. At the end of his first year, he spoke of a life not having been in vain, of "benefits that would accrue like an overflowing river." [1] In 1900, however,

[1] "Since our assuming the presidency our time day and night has been our fellow craftsmen's to command, we can only add that whatever avenue of life we may find ourselves placed in that the interests and welfare of our brother miners will be our first consideration, and if we can be of service in making life more pleasant by increasing their comfort and happiness and bringing joy and gladness into their homes instead of gloom and sadness . . . our lives will not have been spent in vain. We have met with nothing but kindness and for our successors we could desire nothing better. Trusting that our deliberations will be harmonious and that benefits like an overflowing river will accrue to our fellow miners, I return to your keeping the trust your representatives have placed on me."

at the end of his second year in office, he was strong enough himself to say in public that the union was not strong. After a pardonable pride in the accomplishments of the year, he laid most of the stress on "prudence and business methods." The last penny of market conditions to which the miners were entitled should be insisted upon, but in the same spirit they should accord to the operators those profits to which they were properly entitled. Official acts could only be sustained by public opinion.

This time he had taken the alternative of making such a splendid achievement that it would swamp opposition and had won. He was reëlected by acclamation. Tom Lewis, his bitter enemy, however, was elected as vice-president, a position where he could not only claim whatever credit there was to be had from the administration achievements but could work against Mitchell on a large scale. He did not win the election without a fight, but the opposition to him was not effectively managed. Either Mitchell had completely unbounded confidence in his own ability to repeat his successes or he was blind to the difference of character that separated his own somewhat eager and ready trust in his fellow-men from his vice-president. The two men met the proposal to increase their salaries in a way characteristic to each of them. Mitchell commented:

> "I appreciate the honors that you have done me, and assure you that if advancing my salary was to injure the organization, if one man was to go away from here dissatisfied about the matter, then I say you should put the salary at last year's figures or even make them smaller. . . ."

Tom Lewis said bluntly:

> "I do not believe in the policy of attempting to advance mining rates unless we also advance salaries. It will give the joint convention next week a chance to say that we do not advance our officers' salaries because we did not get an advance ourselves. . . . I am one of the people who if I think my services are worth a certain sum, are going to ask for it, and if I don't get it, I am going on a strike."

The raise was not given. Tom Lewis, however, did not go on strike either.

From this convention the miners went to their third meeting with the operators of the four central competitive states. Here they secured increases approximating twenty per cent, making, in Mitchell's words, "the highest mining price paid since the formation of our organization if not in the history of the mining industry." It covered some hundred thousand miners. But he knew that the function of the union still included the matter of paying as it went. If he was to make the union strong in the outlying fields he must exact the price of no trouble at home. At the same time that he gave notice to the operators who had signed up with the union that the sacredness of contract applied to them—that they had a responsibility to the union and that neglect of it would mean that he would sanction a strike—he told his own men that it was a two-edged sword, that they had to keep the contract themselves or suffer suspension from the organization. He wanted a free hand for the outlying fields without the loss of reputation or expenditure of effort which might follow outlaw strikes at home. It was the statement of a man who saw his immediate situation clearly. In this most important industry—since characterized as the worst-run of all national industries—this young man of twenty-nine took upon himself the assumption of executive power. His modesty and his caution had nothing to do with the burning necessity of making good on the agreement.

With the approval and support of his organization behind him and his policies, a satisfactory agreement in the central field, gains elsewhere—Mitchell, had he been a vain man or one eager for power, might easily have overstepped himself. The greater part of the task was, in his opinion, still before him. Mitchell was no mere "prudent businessman," no matter how much he might so protest at the conference; he was an organizer, the head of a militant union. Three hundred thousand miners (including the anthracite fields) were still unorganized; the strike in the Indian Territory was in progress; West Virginia was still to be unionized; and the great stronghold of the trusts in the coal industry—the anthracite fields of Pennsylvania—were still to be entered. Mitchell left the joint conference for the unorganized fields, devoting himself particularly to the anthracite section. Before 1900 was over, the "prudent businessman" had been transformed into a crusader, a strike leader, and it was in this rôle that public attention and approval were centered on him and his own position in the union rendered impregnable.

V

THE CHALLENGE TO DESPAIR

THERE were still two great sections of the coal fields where the proposition that men should be allowed to organize and bargain together against great combinations of wealth had met with no success. West Virginia, with its thirty-five thousand soft-coal miners, was one of them, and the anthracite field, with one hundred and fifty thousand men, the other. Mitchell chose the latter for the field of his activity. His personal opponent, Vice-President Lewis, went down to West Virginia; perhaps he could do better there than Mitchell had done two years before, in 1898.

It was in this field, in the eastern part of Pennsylvania, that Mitchell was to gain a reputation of being something close to a superman, a Lincoln of the oppressed hard-coal miner. He entered this work with some advantages he had not had when he had attempted to organize West Virginia. Now men knew his name, knew that he had brought success with him, had put back food in houses where there had been none. They knew, in a vague way, of his childhood, not unlike their own. They knew of his sudden rise to a position where newspapers quoted him and political figures associated with him. He was the walking proof that their own hopes might come true. He had given them a dignity and importance in the country which they had never had.

Jack Hawthorne, Pete Henderson and Jack Mitchell.
Spring Valley, Ill. 1888

Mitchell entered the anthracite regions in the fall of 1899 under a mandate of the convention of that year. This was embodied in a resolution brought in by Ben James, one of the three delegates who were the only representatives of the local unions which existed in the region. From the vague wording of the resolution, the motivating impulse behind it was the desire to organize the hard-coal fields to prevent their being used as a means of defeating struggles in bituminous fields. The pioneers in the hard-coal fields, John Fahy and Miles Dougherty, were soft-coal miners.

What Mitchell found in the mine camps of the Susquehanna and Schuylkill valleys was the stuff out of which weaker men than he had made tears. There were as many miners to be organized in the five hundred square miles of Pennsylvania as had walked out in response to the call of 1897, more than twice as many as were enrolled under the banner of the union. These men worked and lived under conditions which could not be adequately described by all the vivid phrases used to picture the hardships and injustice in the lot of the soft-coal miner. Starvation wages and inhumanly long hours were not all that kept them helpless; they were kept in a condition of servitude through the system of company-owned houses, enforced purchases in company stores, deductions for company medical services they rarely received. They were bound as serfs to their hard lot. In the anthracite regions, dockage charges were more unjust, the cost of supplies more exorbitant, hours longer, work more uncertain, favoritism more unjust, child labor more cruelly widespread, and the death and accident toll much higher than in the worst period of soft-coal history Their greater misery did not make the anthracite miners more revolutionary, more eager to listen to the gospel of unionism. It made them more hopelessly despairing.

The concentration of the entire industry in five hundred square miles of Pennsylvania territory should have lightened the task of organization in contrast to the scattering of the soft-coal fields throughout twenty-seven states. But within that area were barriers far more insuperable than those of distance. These barriers were harder to mount and overcome than the stark high hills which separated one anthracite camp from another. The miners themselves were divided by racial, national, economic and social differences. The economic line was drawn between the skilled miners who became contractors, and the unskilled "helpers." This line was reënforced by the barriers of race and religion. The skilled men were the older group and English-speaking—English, Welsh, Scotch and Irish. The helpers were the despised new immigrants from Southern and Eastern Europe, grouped together by the English-speaking group as "dagoes" and "hunks," but actually divided into almost twenty cultural groups, each with its own language and church, its own poverty-stricken settlement or "patch" of company houses. The recollection that these foreigners, hated enough for their strange ways, had been imported as strike breakers by the thousands during and after the lost strikes of '75 and '87, was wormwood and gall to the older group. They told Mitchell that these "dagoes" and "hunks" could not be organized; they were ignorant, they were low; they lived like animals. Wasn't it true that five immigrant families lived in company houses discarded as uninhabitable by one "white man's" family? Wasn't it the children of these foreigners who made up the bulk of the army of twenty thousand breaker boys between the ages of ten and sixteen? The English-speaking miners took out their revenge by making it practically impossible for the "dagoes" and "hunks" to become skilled miners, maintaining in a world of economic suffering what social prestige they could. The foreigners hated them almost as much as they hated the foreman and the vast, impersonal machinery which was crushing them.

The English-speaking skilled miners themselves offered very little hope. There were so many living participants in past failures who reminded each other and the others of the futility of organization. Had not the leader of their first union, back in 1850, absconded with the organization funds after a lost strike? Would not one field scab on another, as they had done during the "Long Strike" of 1875? They had struck twelve years before against a ten per cent reduction, and had ended up by being forced to take a twenty per cent cut. What could John Mitchell, stranger and soft-coal miner, accomplish in the stronghold of monopoly, when John Siney, the ablest of the early leaders, had failed, and had, to use their own language, been "crucified" by the men themselves? Old hatreds, old distrusts, old fears were strong.

The employers were cognizant of this division in the ranks of the men. But, unlike the operators in the soft-coal field, who were many, who were divided and scattered, the anthracite operators were united into one of the most powerful monopolies in the country. Ownership and control was concentrated in the hands of seven coal-carrying railroads, which in turn were combined into two groups, one controlled by the Morgan financial powers and the other by the Vanderbilt combine. The coal miners had seen the terrible defeat in 1892 and again in 1900 of the best organized workers in the country, the steel workers, by the steel trust. What chance had one hundred and fifty thousand unorganized miners, half of them foreigners, against the coal monopoly?

Mitchell brought something new into the situation to counteract and overcome racial hatreds and distrusts, the tradition of defeat and sectional treachery, the power of the coal monopoly. He was the living embodiment of the success of the soft-coal miners. The fact that this was an argument used against him by the anthracite operators strengthened rather than weakened his cause with the miners. More than that, he brought them the promise that the bituminous miners would aid them in a very real way during any strike. He could bring them the plan of the union, reënforced by his own standing and the trust men had in his word, that this time for once the three districts in the anthracite would not only strike together but would stay together until the very end. Finally, his personality, and the legend of this successful "man from the west" who was to lead them to victory, made their mark. The English-speaking miners saw in him one of their own men. To the foreigners, this sober, slender young man, quiet, sympathetic and serious, dressed in the long miner's coat worn by the westerners, and buttoned up to a straight clerical collar, seemed much like their own priests. They attached to him the reverence they had for their priests. The warm, magnetic personality, the tolerant understanding which had won over delegates at miners' conventions and subordinate officers, made its own appeal. Here was a leader.

Finally, Mitchell brought to them a new policy, a new plan for unity. Here was the problem: Mitchell could throw his lot with the skilled and English-speaking miners. For such a step he could be sure of praise from them, from some of the leaders in the craft unions. It was not enough. Action based on such exclusive groupings had failed too often in the anthracite field for the new immigrants, whose assistance was absolutely necessary, to give it their support. If the aliens were not to be fought, they had to be gathered in. He called to his aid, for work among their own group, the few leaders among the English-speaking miners, and then set himself to the task of finding and developing leaders from among the foreigners. He went out into the field searching for such men amongst the miners themselves, interviewing every possible candidate, checking back on the man's references in the local foreign-language associations, lodges and clergy. When he found able men, he gave them office on a standing with the English-speaking organizers. He selected men who understood their own language following, but who, even more, understood the meaning of unionism. In the minds of many of them, the cause of unionism became identified with the figure of Mitchell.

To all of his aides in the great cause, Mitchell brought not only a new technique of organizing, but a new spirit. As he went from one district to another, one mine town and pit-head to another, addressing them in halls and in the open fields, talking to them from the pulpits of their churches, visiting their homes, he emphasized over and over a simple but much-needed doctrine. He appealed to them to forget their racial antagonisms and all lines which divided them. The foreigner was no longer to be despised; he too was a miner and fellow-sufferer. "The coal you dig," he pleaded, "isn't Slavish or Polish or Irish coal, it's coal." Coal was the leveler. It brought its own terrible equality. He asked the English-speaking miners, in particular, "who failed to treat these fellow-workers who did not speak the English language . . . with consideration . . . who called them Hunks . . . that they should stop that practice and give these foreigners as much consideration as they demanded for themselves . . . to call them Mr. So and So or to learn their first names. . . ." The miners of all groups and languages came to hear him. They listened for hours on end in the open fields as Mitchell spoke in his slow unoratorical fashion to them. They could not all understand the words but they caught the spirit long before their own language interpreters, one after another, translated what he had said. A new day, the miners felt, was coming.

Mitchell looked around for other help. In the old days in Spring Valley, Father Power and other priests had exercised great influence over both miners and operators. In the anthracite region he knew the influence of the priests and foreign language editors was even stronger. He met with every priest and clergyman throughout the district, pleading the cause of the union. He pointed out what they already knew, that crime and lawlessness followed in the wake of the wretched poverty of the district, that the new generation was destroyed in its childhood. The union which he represented, and he himself, embodied a way out of poverty and lawlessness. He was most successful with the Catholic clergy which depended for support on and had most intimate contact with the masses. This non-Catholic leader, so often mistaken for a priest, won over Bishop Hoban in Scranton, made a strong ally of Father J. J. Curran of Wilkesbarre, himself once a breaker boy, and hitherto antagonistic to the union, and, after a public debate, converted Father Phillips of Hazleton. These three men had great influence not only with the miners but with the operators. The Protestant clergy, too, despite the dependence of their churches for support from the employing and tradesmen classes, rallied to the miners. No man was overlooked. When a local union leader complained bitterly against a Lutheran minister who had advised his parishioners against the union, and suggested darkly that "something should be done," Mitchell made a friendly call on the preacher. He did not attempt directly to change his point of view, but later, during the strike, this minister did much to aid the strikers.

Mitchell went further in his program of winning "public opinion." No group or agency in the anthracite field or outside of it was left untouched. John J. Loftus, a Scranton pharmacist with a wide acquaintance and political connections, who had become his loyal friend, after some effort, induced the Elks' organization to give Mitchell a hearing. They intended it as an opportunity of giving him a friendly warning to keep his hands off that part of the country, but before the meeting was over he had won their support, important since they constituted the tradespeople and general citizenry of the section who might be asked to extend credit to the strikers. His obvious earnestness, his utter lack of demagogy, appealed to them. His argument that general prosperity could be built only on the foundation of a well-paid working population with money to spend aimed at their self-interest.

At the shabby, one-room headquarters in the old Valley Hotel in Hazleton, purposely selected both for the sake of economy and so that the miners could feel free to come as they wished, Mitchell sat in his shirt sleeves, planning the work and talking with his people. The work became an unremitting round of bolstering up timid hopes, flattering hurt vanities, appealing to what unselfishness and what courage had been left over from all the years during which the anthracite industrialists had relentlessly eliminated the most fit from their enterprises. After five months of work in the field, there were still only ten thousand union members out of a total of one hundred and fifty thousand workers. He left the field to report this fact to the national field. When he came back, the bituminous miners had received a twenty per cent increase. It was an open advertisement to the doubters in the anthracite field that the thing could be done. Some of them had taken so many risks and lost, that now they wanted water-tight, fool-proof, completely guaranteed assurance that there would be no more failure. Others were so eager of results that they struck immediately, but in the old way, closing down only a few scattered collieries.

In August of 1900, just as the presidential contest between

McKinley and Bryan was getting seriously under way, Mitchell called a convention of the three anthracite districts to meet in Scranton. He had been in the region almost a year. There had been a great increase in the union enrollment. Would the operators pay any attention to him? Would the bulk of the men really come out? Would the politicians do anything to avoid having the issue of social injustice figure in the campaign? Would the people in New York City and the other cities on the eastern seaboard who heated their homes with anthracite coal be with the miners, or against them?

The convention worked out demands. They wanted wage increases. They wanted to be paid on a piecework basis of twenty-two hundred and forty pounds to the ton instead of the "long" ton (twenty-seven hundred to four thousand pounds) basis. They wanted relief from the system by which the operators were free to dock their wages for almost any amount of what they chose to call unmarketable coal. These were the main things. They had received no increase since 1880, twenty years. The working days had been cut down to one hundred and ninety a year. They wanted relief badly. What could Mitchell promise them toward it?

For the moment it seemed to be very little. An appeal to the operators to meet with the district officers and himself was, in spite of the prestige he was supposed to have, completely disregarded. The National Executive Board gave him authority to call the whole anthracite field out on strike within ten days. Did the operators pay any attention to that threat? Into their intense silence he sent a wire suggesting arbitration of the issues. It was ignored as all his other communications had been. He was being shown his place in the East.

He knew that the operators could, on the face of things, very well afford to ignore him. They were the great financial leaders of the country, men who had not only seen the railroad and steel workers crushed by their colleagues in finance, but knew exactly how it had been done. There was an opinion current in their circles that men had their price. In the face of this attitude there was only one thing for Mitchell to do—call the strike and take a chance that enough men would come out to show both the others and the operators that the union had at least that power. The only other alternative, of waiting until the official enrollment had grown more impressive, was a matter of impossible years.

On September 17, 1900, the strike call went into effect. In the mining camps, at the forks of the creek, in the hills of the Schuylkill and Lehigh Valley, in Scranton, Wilkesbarre, Pittston, in the flat cities of the north, the whistles blew for work that morning. The miners woke and listened, stood up and got dressed, and then, in spite of all their need for every cent they could earn, some hundred and twelve thousand men stayed at home, and heard the whistles blow again, and still stayed at home, and hoped and prayed that other men throughout the region whom they had never seen and never would see had also determined to stand fast. It was a matter of three days before the news got to the remote camps that a real strike had begun at last, that there was a good chance that this time their courage would not have to be hidden away, that they would not have to plead with the bosses to forgive them their trespasses.

Against Morgan, Vanderbilt and the other financiers in New York it was not hard to raise a righteous fist. There were, however, a few independent operators who lived in the fields, who prided themselves on their close personal relationship with the men who worked for them, their workers' clubhouses, the baskets their wives took to the sick. They felt they were not unreasonable. John Markle, one of the most important of them, was petitioned by his twenty-four hundred employees to grant some demands. He promised an answer and offered arbitration. It would have set a dangerous policy of defection. A meeting of these men was held at the Jeddo School House two days after the strike started. Mitchell invited himself along. Against Markle's offer of arbitration for his men, against a warning by the much beloved Father Phillips that they should not break their unwritten contract, Mitchell could only point out what he had learned, that wages and conditions were not a local and individual issue, that terms now won by the Markle men could not be enforced long unless the same wages prevailed throughout the whole region, unless the union was established. Arbitration was useless unless it included the railroad corporations which controlled more than three-quarters of the coal in this section. He pleaded with them in the name of their struggling fellow-workmen who did not enjoy the advantage of a philanthropic employer. News of the strike's success was coming in every hour; the Markle men decided to stay on strike. Before it was over, Father Phillips had been won over to the union cause. The great danger of local settlements had been averted.

As soon as it was clear to the country that the miners really were on strike, and stories of the actual misery they were trying to alleviate were told, the people who were interested in electing a president on the proposition that he had brought and would bring a full dinner pail with him, began to stir. Those who were interested in defeating a Republican president on the ground that he had not brought a full dinner pail also began to stir. The interest of Hanna and McKinley in the Ohio miners, the fact that Bryan had seemed to embody to the workers four years ago some fulfillment of their hopes, were remembered.

Marcus Alonzo Hanna, who was to play an important part both in the 1900 and 1902 strikes, was at this time considered not only boss of the Republican Party, creator of the "stand pat" policy, but spokesman for the financial powers of the country. Originally belonging to the industrial captains, his fortune made on forfeit during Civil War days, he seemed to have and to retain a human sympathy for men who worked and fought hard. He had bargained with the union in Ohio ten years before other operators had been forced to it. Some of his most valuable lieutenants—the radicals called them his labor henchmen—were miners' leaders. Since he had gotten into national politics he had learned that the men really running the country were not the industrialists like himself but their financiers. They, as absentee landlords, were even less inclined to deal with the unions than the industrialists. When he became the dominant force in the National Civic Federation, he used his position to impress these men with his conviction that unions were a necessary evil and something to be accepted speedily and intelligently. The Federation, at that time, had the hearty hatred of the manufacturers' associations. Hanna could be counted on for assistance.

There was only one of the four questions left. The bulk of the men had come out. The operators had paid no attention so far. The politicians were beginning to consider the effect that an issue of social injustice might have in the campaign. What remained in doubt was whether the citizenry and consumers would feel with the miners or against them. Mitchell wanted their support. He preferred to take the criticism of the radicals in his organization that he was a mild conservative rather than to alienate middle-class sympathy. One of his first appeals to the miners was to avoid violence. There was to be no opportunity to repeat the horrors of the Lattimer massacre of 1897, when forty of the foreign-born miners, marching from camp to camp, an American flag in the hands of their leader, were shot down and nineteen of them killed. The picture of an almost priest-like young gentleman, coming out of the West to stand up in behalf of humble men against all the powers that seemed to be controlling industry against the will of the common people and bringing with him peace rather than revolution, captured the imagination of both newspaper men and reading public.

When the operators, in public pronouncements, raised fine points about whether or not the union represented their workers, even the conservative *New York Times* insisted that that was beside the real point of the validity of the strikers' claims. The suffering of the miners' families and their devotion to the union made the hair-splitting of their opponents ridiculous. A priest in the region was quoted as saying that the men were now more loyal to the union than to the church. The loyalty extended to their wives and children. A story was told of an old woman who put all her husband's clothes into the washtub to guard against any possible yielding on his part to the temptation to return to work. Privation and want threatened the strikers. There was not always enough bread, and butter was a luxury unknown. Actual starvation was warded off by the helpful extension of credit by local tradesmen, now thoroughly with the miners. Breaks in the ranks of the strikers were unknown.

The usual tales of strike violence failed of appeal. The press was agreed that here was mass action and much passion, but violence was unheard of. "The quiet in the mining hamlets (is) so intense, so unnatural as to arouse the feeling of uneasiness." The day after the strike call, the pledge of temperance had been taken by many of the strikers and administered by the priests at the request of the union. Disorders were fewer than in peace time. Not even the cry of destruction of property could be raised, since the union had not called out the pumpmen and engineers, and the mines had not been flooded.

Then the operators pronounced it a plot on the part of soft-coal operators in conspiracy with the union. The outspoken sympathy of the bituminous operators for the cause of the strikers made that plausible until it was pointed out that even prior to the strike, due to the great industrial activity, the

Miners' Homes, Showing the Mine Shaft in the Distance

demand for soft coal was greater than the supply, and that, in ordinary times the two products did not directly compete with each other. The press declared the issue an evasion of the strikers' claims of injustice.

Then the operators, who themselves were high in political councils, insisted, with great indignation, that the strike had been "pulled for political purposes." It was a matter of general knowledge that Senator Hanna was using every means he could to bring about a settlement. Opponents of the union accused it of conniving with now the Democratic and now the Republican Party, the accusation depending upon the politics of its maker. Mitchell's method of meeting such accusations when he was directly confronted with them is illustrated in a statement, issued by Miss Morris: A certain William V. Webber of the "German-American McKinley Roosevelt League" was quoted as saying that the anthracite coal strike had been planned by Mr. Johnson of the National Democratic Committee, the Mayor of Indianapolis and John Mitchell. Mitchell commented that "he was reluctant to believe that Mr. Webber made the charge and stated positively that the story was untrue; that he did not know Mr. Johnson and had never held a conversation with Mr. Taggart in his life . . . that any man who would inaugurate a strike in the interest of any political organization would be deserving of the severest punishment which could be meted out to him, and that any persons who made political capital out of the suffering and misery of the men, women and children of the anthracite region would be unworthy of the privilege of citizenship." In all his life, Mitchell never attacked a man, either in defense of himself or of his ideas. Every accusation made against the miners' cause was turned into an argument for it, partly by virtue of its own righteousness, and the unfair nature of the attacks, and partly by the temper of Mitchell's replies.

It was said that the strike was a fight for personal prestige and union recognition only. Mitchell replied that the matter of union recognition would not stand in the way of a fair settlement.

Early in October there were signs of the end of the strike. It was getting too close to election time. The operators made overtures to the strikers by posting notices of a ten per cent increase for those men who would return to the mines immediately. No mention of the other demands was made, no promise or guarantee as to the length of time to be covered by this increase was given, and no method of adjusting grievances included. The union decided to continue the strike and hardly a man broke ranks.

During all of this time, Hanna was busy. When he found the operators firm against any discussion, he went over their heads to J. P. Morgan and arranged an interview for Mitchell. The desire for political victory commingled with admiration for Mitchell's personality. No public announcement of the result of this interview was made, but the railroad presidents received their orders.

On October 3, 1900, eight hundred delegates, representing every colliery in the region, met in convention and decided on certain compromise terms: a ten per cent increase until April, 1901; a reduction in the price of powder of almost one hundred per cent; the abolition of the sliding scale of wages (by which wage rates fluctuated with the market price of coal) in the southern and western fields; and finally, recognition of workers' committees in the adjustment of grievances.

The operators gave no sign of awareness of the convention, but a week later notices were posted in and about the mines embodying these compromise terms. The notices were tacit acknowledgment of the strength of the miners, and were accepted as such by them. It was good enough. It was victory. The country at large considered it a victory in the struggle against industrial feudalism. On October 29, 1900, an order was issued by the national and district officers, announcing the end of the strike.

The miners entered the strike, unorganized, helpless, uncertain, ignored by the operators, divided amongst themselves. At the end of six weeks they emerged as a power to be dealt with, a group with a definite status in the industry and in the community. They had not won formal recognition, but in effect their union had been recognized. In the national union, they now represented almost as large a group as all the bituminous miners. The northern district alone was almost as large as the most powerful, militant district in the soft-coal section, Illinois. Mitchell announced that the strike "stands out in bold relief as the most remarkable contest between labor and capital in the industrial history of our nation."

Mitchell, as leader of that strike, was, to the men and to the public, the personification of the new power of labor and the new attitude toward unionism not only in the anthracite region but throughout the country. October 29th was voted a holiday by the miners and called Mitchell Day. The press spread itself in his praise; the miners spread a worship of him as a Moses who had wrestled with their Pharaohs, a saviour, a Lincoln. Before he left the region he was loaded down with gifts and memorials. The twenty thousand breaker boys, pathetically proud, presented him with a gold medallion which touched his sentimental pride in them; their lot, as sad as his own boyhood, had made his work in the anthracite field, a mission, a crusade. Local unions, district executive boards, foreign-language associations, Elks, all fêted him and marched in his honor.

This was his first great fighting triumph. Other victories had been won by his powers of negotiation, by competent executive control of other leaders. His triumph over attacks on him outside the union, the new dimensions of his leadership, did not save Mitchell in the years between 1900 and 1902 from attacks within the union both on his policy and on his personal integrity. There was little enough time for him to rest on his new laurels.

The opposition to Mitchell at this time was mainly that led by Tom Lewis, whose personal ambition and jealousy led to many attempts to undermine Mitchell's prestige. But there was in addition a small but growing opposition, led by men whose concepts of the function and methods of unionism differed fundamentally from the policies which Mitchell represented. To them—whether from time to time they enrolled in the Socialist, the Socialist-Labor, or the I. W. W. ranks—he was too conservative, too concerned with immediate results and not enough with the revolutionary future of labor, too willing to be on amicable terms with capitalists. His membership, together with that of other international union presidents, in the National Civic Federation did not appeal to them. The Civic Federation, formed in 1900, was composed of capitalists, representatives of organized labor, and representatives of the public. One of its aims, set forth rather patronizingly in its preamble, was to "show that organized labor cannot be destroyed without the debasement of the masses . . . that organized labor can be led to correct its errors . . . that capital can be taught the practicability of securing industrial peace in accordance with business methods . . . that the twin foes of industrial peace are the anti-union employers and the Socialists, and that the former are unconsciously promoting that class hatred which the latter boldly advocates." To the leaders of the radical minority in the union, this seeming advocacy of organized labor was more to be feared than the open enmity of certain capitalists. The leaders of this minority group, just beginning to be articulate at this time, were practically all Illinois miners—Duncan McDonald, an eloquent orator, Frank J. Hayes, who was writing poetry at this time but whose engaging personality won him the presidency of the union a dozen years later, Adolph Germer, who was active in the socialist party. Occasionally one of them, like John H. Walker, supported Mitchell out of personal loyalty and a conviction that his achievements were sounder than the pronouncements of more radical leaders. Their opposition was important then, not in itself, but in the use which Lewis might, in his frequent bids for their support, make of them.

The Lewis-Dolan group found it a little difficult to discover weaknesses in Mitchell's armor. The organization had grown in the years between 1898 and 1902 to over two hundred thousand from less than forty thousand. Every month new territories were brought into the effective ranks. Every convention was a testimonial to Mitchell's leadership. In 1901 the anthracite miners, out of their meager earnings, made a collection to build a home for Mitchell. He insisted that it be converted into a fund for a memorial to the sturdy alien strikers who had died in the bloody Lattimer massacre. Mitchell's salary of $1,200 a year, small compared with the salaries of other labor officials holding positions in unions one-tenth the size of the miners' organizations, was, he said, sufficient for his needs. It was—and he pointed that out—far more than the vast majority of miners earned; he permitted no resolutions for its increase to be considered for many years.

The Lewis bid for radical support by his resolution that National Executive Board members be elected by district conventions instead of by the national convention, was belied by his opposition to the proposal of Mitchell that national officers be elected by direct referendum, instead of by the convention delegates. Mitchell's proposal was not carried until 1902. Lewis' bid for the support of the foreign groups in a resolution for the appointment of foreign organizers and the introduction of foreign language sections in the *Journal* went by unnoticed in the acclaim of the foreign-born for the man who had brought them victory in the anthracite fields.

Radical propaganda for industrial unionism was made to look futile and feeble beside the quiet practical achievement of Mitchell, in gathering into the ranks of the union not only the unskilled unorganized miners at one extreme, but the highly skilled, well organized engineers, pumpmen, firemen, carpenters, and blacksmiths who worked in and about the mines. His success drew the protest of the craft unions which had jurisdiction over each of these separate branches, and finally, of the American Federation of Labor of which he was now second vice-president of labor. He was rebuked, in 1901, by Mr. Gompers: "Surely the lessons of previous efforts in the organization of labor have been sufficiently costly to our movement and to the workers generally to bid us have a care lest we founder upon the same shoals of industrial concentration. The Knights of Labor endeavored to cover all branches of industry in one, as they termed it, 'comprehensive' organization. . . . There was apparent success for a while, until disadvantages broke out at every point and disrupted the whole. . . ." Finally, after considerable correspondence, the 1901 convention of the Federation in Scranton—the scene of his triumph over the anthracite operators—gave Mitchell and the miners' union the right to include these crafts, which formed, as Mitchell pointed out, but three per cent of the total force but which, if they chose to strike, could throw the remaining ninety-seven per cent out of work. The permissive declaration, however, called for a maintenance of the principle of craft autonomy "in so far as it is consistent with the varying phases and traditions in industry."

It was, indeed, not until the 1902 convention of the miners that an issue was found on which Mitchell might be attacked. Between annual elections Mitchell insisted always that there was no internal conflict. In 1902 the fight was made on the very delicate point of Mitchell's personal integrity as well as his judgment of men. The national secretary-treasurer of the organization was W. C. Pearce, who had held that office since Ratchford's first term in 1897. He was an Ohio man, reputed to be none too friendly to Mitchell. In the spring of 1900, Pearce's secretary, a Miss Mollie Meredith, reported to Mitchell a shortage in his stamp accounts and a general laxness in his accounting methods. Mitchell showed unexpected concern over the matter and his statement to Miss Meredith, "I wish I had never to hear of such a matter," was given its own interpretation by his opponents at a later time. He sought the counsel of a national official, W. D. Van Horn, stationed nearby at Terre Haute. Van Horn's solution was simple; the man must resign. He could not understand Mitchell's extreme concern, his astonishment and his tears, over the defection of a fellow official. Pearce, confronted, admitted a shortage of a hundred dollars, but put the appealing defense that his own funds had year after year been steadily depleted by his assistance to the many destitute miners who called at the national office. He was a broken and pitiful figure and Mitchell hesitated. But when Pearce suggested that the matter be fixed up quietly and that he be allowed to stay on until summer when he expected a political appointment, Mitchell's sympathetic mood vanished. Pearce was forced to resign, and the official pretext was ill health. Miss Meredith was rewarded for her services to the union, which gained a more efficient secretary in the person of William B. Wilson of Pennsylvania, an old and tried official. The Pearce matter was to all appearances dropped. Pennsylvania now had a representative in the national office, even though he came from the Clearfield section, and not from the Pittsburgh district Dolan controlled. Mitchell had insisted on the policy of official honesty.

His honesty did not go unchallenged, however. A short time before the 1902 convention, Miss Meredith, no longer in the employ of the national office, appeared at a meeting of the Ohio district and charged Mitchell and Wilson with glossing over the Pearce default by failing to mention rake-offs Pearce obtained on printing contracts. She repeated Mitchell's exclamation to her. She had undoubtedly been brought to the convention by Mitchell's opponents, but her story sounded plausible. She was a handsome woman who appeared in the rôle of an employee abused for her honesty, and the Lewis men were in control of the district convention. It voted to pay her expenses to the national convention at Indianapolis and demanded that she be heard there. The discussion of her charges took three days of the national convention's expensive time. The press found good material in the spectacle of a young woman, good-looking, becomingly dressed in the height of the fashion, standing out seemingly alone against the powerful president of a powerful organization. Many of the delegates were inclined to think that there might be fire where the smoke emanated from such a source. Mitchell was embarrassed.

At that time miners' conventions were held in an old, badly underheated hall. It was a cold January day and most of the delegates wore their coats indoors, but Mitchell's face when he arose to answer Miss Meredith was, according to the reporter for the *Indianapolis News*, bathed in perspiration. He spoke with great emotion. He denied that he had failed to report the full shortage to the Executive Board. He denied any plan to cover Pearce's misdeed, pointing out that he could have made no personal gain by such an act. He protested against Miss Meredith's interpretation of his expression of regret. He did not refrain from pointing out that she had not made the charges until after her own dismissal for inefficiency.

The convention was in an uproar. Resolutions were introduced, some censuring Miss Meredith, others exonerating Mitchell immediately, others calling for a vote. An investigation committee was finally selected, including many of Mitchell's opponents, and Mitchell refused to have his name considered for reëlection until it had been cleared. When the committee brought in its report, completely clearing Mitchell and Wilson, the delegates arose and cheered. Rumors became more insistent that Miss Meredith had been employed either by the operators, or through them, by the anti-Mitchell men in Ohio. The delegates sang the "Mitchell song," composed in his honor by a miner-singer in the anthracite region, over and over and ended the last session with it.

The whole matter touched Mitchell on the raw. He knew his own weaknesses. The boy who had been scorned and mocked, who had been for many years a wanderer among strangers, still remained turned in on himself. He was not toughened to the politics of an organization. As he met men in other classes of life, he realized how inadequate his own education had been. That too made him reticent. When people appealed to him in the name of friendship they met a fund of goodness and simple decency. Some found the very sincerity of this quality disarming; others found in it an opportunity. It hurt Mitchell to be accused of the least treason to the union which was his life and family and religion. This was not the last time he was to be accused. It hurt him more to have any of those people who really were his friends accused or made to suffer through him. In the midst of the anthracite strike of 1900, one of the Scranton papers published serious insinuations about his relations to Miss Morris. He promptly brought a libel suit against the editor and forced him to make a public retraction. One of Mitchell's friends tells that at another time Mitchell suddenly left him to follow into a saloon and punish a miner from Ohio of whom he had heard that he had made a similar accusation.

Another serious charge was made in 1901. The previous year the steel workers' union made a last desperate fight to stay in the mills of the newly formed United States Steel Corporation. T. J. Schaffer, President of the Amalgamated Steel and Iron Workers' Union, claimed to have received a promise from Mitchell that he would call out the coal miners in a sympathetic strike, on which promise the steel workers had counted before they took the risk of the strike. Mitchell denied this in full. He pointed out that both Gompers and he had not approved of Schaffer's policy of breaking the agreements in the unionized mills in support of strikes in unorganized mills. Publication of the correspondence seemed to clear Mitchell completely of any broken promise or hostile activity. He was not present at the meeting of his National Executive Board when it endorsed an appeal to its own locals for aid, and a resolution to the American Federation of Labor requesting President Gompers to call a mass meeting to consider means of aiding the steel strikers. The reason for his absence was that he had suddenly left for Spring Valley to attend the funeral of one of his children. He had never known the boy very well. The last few years had kept him almost incessantly busy. He could not stay on to know the other children better. Hardly was the funeral over than he had to leave for a convention of the anthracite miners. The Schaffer attack, the first charge flung at him from outside his own union, together with his emotional state, made him abnormally sensitive about it.

All of these attacks made Mitchell realize that he was responsible not only to himself and to his people but to all those throughout the country who placed their hopes in him as the bringer of a new era in social justice. Any mis-step, of diplomacy as well as of economic program, would lead to bitter factional disruption. The increased complexity of the tasks before him, and the power concentrated in his office by the constitution of his union, enlarged the opportunities for such a mis-step. Mitchell left the convention of 1902 and the joint conference which followed in depressed spirits and health. He had been vindicated in the convention, he had won gains for the miners in the conference. But he was said to be tortured with pain—a hernia resulting from overstrain in his days as a miner. His wanderings prevented any domestic existence. His own children did not know him. He disliked the internal conflict in his organization. The immediate tasks before him were too pressing to permit him to dwell on his emotions.

The two great threats to a unified organization covering

all the coal fields in the country remained. After the 1902 convention Mitchell went back to the anthracite fields, where one test of strength had made the men willing to take the great risk of still another test. Tom Lewis, whose reëlection to the vice-presidency he had not opposed, went to West Virginia. It was a double attack The work of unionization was never finished.

In the anthracite field, Mitchell found that for lack of a union with disciplinary powers and unified control there had been more local stoppages in the months since the victory of 1900 than in the preceding year. Some of the foreign miners were telling their foremen, "You not my boss, Johnny Mitchell my boss." The operators liked to think that in 1900 they had given in only because of the emergency of their favorite political party. They were in no amicable mood. An elaborate system of spying and blacklisting had been established. The workers had the opportunity of seeing stockades built around some of the mines, huge depots established for the storage of a strike reserve of coal.

In September, 1901, McKinley was shot by a foreigner, and Roosevelt, whom the financial powers did not like and whom they had helped to kick upward into the vice-presidency to get him out of New York State, succeeded him. It is said that some of the foreigners in the anthracite region, when they heard of the death of a president, went into mourning for Mitchell. He was the only president they knew.

In 1901, the operators had met an attempt on Mitchell's part to call a joint conference with a silence equal to that of the previous year. They had simply posted in their mines a notice that the scale would be continued another year. The convention of the anthracite miners was restless that year. Through Mark Hanna's efforts, Mitchell and the anthracite district officials met President Thomas of the Erie Railroad, in New York City. Nothing was gained there except a repetition of the announcement that the scale would be continued until April, 1902. Mitchell said later that the Union representatives had "the hope, if not the anticipation, that the union would be ultimately recognized." The organ of the Socialist Labor Party, *The Weekly People*, sneered: "Mitchell's mission is to prove to the companies that his union is dollars and cents in their pockets." From the other side, the *Coal Trade Journal* commented knowingly: "Mr. Mitchell's invitation for a conference will not receive any more attention now than it has in the past. He will be received however by the operators and treated cordially. He will be given a satisfactory answer to questions of minor grievances and promises of a conference at a future date, say a year hence. That will be satisfactory to Mr. Mitchell and he will then prevail upon the members of the Executive Board to accept the offer of the operators. One year hence there may be no President Mitchell." Mitchell threw his weight with those who found the situation unfavorable to another strike.

The men kept their confidence in his leadership. They did not expect to own the earth in a day. No matter how much he might say, "Relations of labor and capital are purely a business proposition. There is no sentiment in it. Peace is better for both," the newspaper men found the air on Mitchell Day heavy with an almost religious devotion as the miners marched in honor of their union and its president. On his side, he too was touched. Their hardly vocal desires went out to him, and for a moment, under the crush and monotony of the daily routine, he saw himself as they did, crusader and saviour. The breaker boys were grouped before him. He said of them: ". . . these boys in the hard-coal regions are men. They have the bodies and faces of boys but they came to meetings where I spoke and stood as still as the men and listened for every word. I was shocked and amazed . . . as I saw those eager eyes peering at me from eager little faces; the fight had a new meaning for me; I felt that I was fighting for the boys, fighting a battle for innocent childhood. . . ."

In 1902 the anthracite leaders pressed for action. The National Executive Board authorized a joint conference, and, in the event of its failure, a general strike in the whole anthracite field. It did not want to break contracts in the bituminous field, but did pass a resolution giving the board power to call sectional or national suspensions of bituminous miners in case any soft coal was shipped into anthracite markets. The friendly bituminous operators took this as a gesture of warning. The radical element in the union took it as a complete authorization for a nation-wide strike. Antagonistic employers everywhere pointed it out as an indication of the slight importance the union attached to its contracts.

When Mitchell reopened attempts at negotiation with the operators, it was evident that the situation had somewhat changed from the fact that his letters were not completely ignored. Mitchell's invitation of February 14, 1902, to a joint conference on March 12th was promptly, though unsatisfactorily answered by all the operators. The most important of these railroad president coal-operators was George F. Baer of the Philadelphia and Reading Railroad, which, in combination with subsidiary roads, controlled fifty per cent of all anthracite production. The public utterances of these two men, John Mitchell, young, uneducated president of a union in a day when unions were regarded as outlaw organizations, and Baer, his senior by a generation, well-trained in the law, bearded and patronizingly benevolent, religious and powerful, was to form one of the interesting phases of conflict between two opposing forces. The shrewd business head was to prove himself a far less apt judge of public opinion than the union leader, uncertain of the endurance power of his forces.

Baer declined Mitchell's invitation to the conference of March 12th, pointing out that a uniform scale was not practical, that Mitchell's proposal to "disturb the scale" showed his lack of familiarity with the anthracite industry, that complaints as to specific grievances from their own men would always be received by the company, but that these men were "well satisfied." That discipline "which was essential to the conduct of all business" had been disturbed by the union, with a resultant decrease in production, and "there can not be two masters in the management of business." The adjustment of wages and conditions did not "call for the intervention of the organization which you, Mr. Mitchell, represent," although he was willing to "concede" that certain "honest" efforts of labor organizations toward better welfare and working conditions were commendable.

Mr. Thomas, of the Erie Railroad, who had met with

Mitchell in 1901, denied a verbal arrangement to meet with the union in 1902, and added that nothing that had occurred in the intervening year had raised the confidence of the operators in the union. The other railroad presidents replied in much the same way.

When the miners' convention took place in Shamokin, March 18th to 24th, 1902, the miners were tired of delays and dodges from the operators, and it took all of Mitchell's conciliatory powers to keep them from the immediate declaration of a strike. In the midst of the convention, he again sent out telegrams to the operators inviting them to a joint conference. Mr. Baer's reply, the only one on record, read: "Always willing to meet our employees. . . . I had hoped my letter clearly expressed our views."

The miners proceeded to a formulation of their demands which included, in addition to a twenty per cent increase, uniformity in the system of weighing coal, the eight-hour day and similar demands, the important provision that the union be recognized and that all discharged union men be reinstated. On Mitchell's advice, the convention again appealed to the National Civic Federation to mediate for them. It also decided that if, after April 1st, no agreement had been reached, the men would work only three days a week. The engineers and pumpmen, however, were not to be called out unless absolutely necessary, so that the mines would not be flooded.

Mitchell's counsel of peace again prevailed and again attempts at negotiation were begun by the National Civic Federation. The operators actually did meet with the miners, but the net result was a promise from Mr. Thomas, spokesman of the operators, that union men would not be discriminated against. On the other hand, he demanded that union men must not refuse to work with non-union miners, and that the quality of work must not deteriorate nor was the output to be restricted. The upshot of it all was that "after discussing at great length the general relations of labor and capital the conference adjourned for thirty days." The second conference brought no further results.

At these conferences Mitchell, in his rôle of advocate of peace, offered several compromises and made "every attempt consistent with the preservation of dignity" to avert the strike, expressing, for the miners, a willingness to settle on a ten per cent increase and a nine-hour day. Mitchell justified this policy a year later when he wrote in his book: "This policy was not dictated by a fear of losing the strike but in order to avert the terrible suffering which it was clearly foreseen would inevitably result from the desperate conflict. This policy was, however, mistaken for weakness and each attempt at conciliation increased the obduracy of the railway presidents." Peacemaking had its disadvantages. Not a single additional point was conceded by the operators.

April 1st, the date of the expiration of the 1901 arrangement, had come and gone. On May 7th, the District Executive Committee, "with a lingering hope of averting the impending strike," sent telegrams to the operators, reiterating their willingness to submit their demands to a committee of five persons, selected by the National Civic Federation, the decision of the committee to be binding for a year. Or, if that was unsatisfactory, they suggested a committee consisting of Archbishop Ireland, Bishop Potter and one other person whom the two would select. The miners went further: if, after investigation, such a committee decided that the wages received by them were sufficient to enable them to live, maintain and educate their families in a manner "comformable to American standards," the miners would withdraw their claims. The operators replied that there was nothing further to discuss. Baer was a little more explicit. Anthracite mining "is a business and not a religious, sentimental or academic proposition." He would not, as the responsible head of a business with its duly constituted management, even if he could "delegate this business management to even so highly respectable a body as the Civic Federation nor can I call to my aid as experts in the mixed problem of business and philanthropy the eminent prelates you have named."

There was now no retreat. Mitchell said, "We have spoken our last word and the time for action has arrived." On the 12th of May, 1902, one hundred and forty thousand miners answered the call for the temporary suspension of work. Would it be permanent or would it fizzle? The declaration of the suspension hardly made a ripple in Wall Street, according to the *New York Times*. John W. ("Bet-you-a-million") Gates, after a conversation with J. P. Morgan, was still willing to bet one hundred to one against the strike, because of the "wisdom of the leaders and the conservative elements amongst the miners." The newspapers reported that, unconcerned, the "coal barons play golf." Mitchell's tactics had indeed convinced them that the strike would fail.

Mark Hanna was still at work on mediation. He went over the heads of the railroad presidents again and cabled to J. P. Morgan in Europe. He tried to bring pressure on the stockholders, getting a committee of them together. He urged Mitchell to name the lowest possible terms he would consider. Mitchell spoke of a five per cent increase, but the railroad presidents were scornful. It is doubtful that the miners knew at the time of Mitchell's offer.

Two days later a convention was called to determine the future of the suspension. Delegates were to come instructed by their locals so that they would take the responsibility for the calling of a strike. Mitchell had made the process of negotiation go as far as he could, and he still hoped for something which seemed to others to be a miracle. On the day of the convention opening, May 14th, he was still urging the National Civic Federation to do what it could. When the delegates asked him for his advice he declared himself opposed to the strike if it could possibly be averted "despite the provocation the miners had received from the operators." A year later in his book he stated that he had hoped it would be averted until the late fall when the men might have struck more effectively.

A delegate who rose to ask the miners to follow the advice of their president, the "Moses who had led them to the Promised Land," was met by the sardonic retort, "Yes, that's just the trouble. It's the Promised Land." On the following day Mitchell insisted that the delegates vote as they had been instructed, and that the discussion be not ended until all had had an opportunity to express their opinions. Two of the three district presidents were against the strike. In the excitement of the news that the companies were already taking steps to discharge all the convention delegates, a telegram from the American Federation of Labor advising against the strike was read but unheeded. The vote to continue the suspension was 461½ in favor and 394¾ against, a narrow margin as strike votes go in the coal fields. For the men voting the way of peace was already the way of defeat. What was later called the greatest strike in American history had been declared.

On May 15, 1902, John Mitchell took up the duties of leadership in a strike which he had opposed, a fight which would reënforce his reputation as "the greatest labor leader in the country." Oddly enough, in the months which were to follow, his opposition to the strike was to constitute one of the strong arguments in his favor with the general public. This was not a struggle instigated by ambitious leaders, but one insisted upon by the rank and file. The *Chicago Record-Herald* stated, a few days later, that the operators foreseeing his opposition, "hoped to smoke him out of the union," and through breaking him, hoped to break the union. They had calculated that the failure to strike would mean the death of the union, but that a strike against Mitchell's advice would lead to his resignation, a leaderless and a lost strike. Whatever the decision, they expected the men to lose. They had not counted on his willingness to accept the verdict of the men. His sincerity in insisting "on the control of the union by its own members" was being tested. John Mitchell entered upon the leadership of the struggle with all the earnestness that had characterized his attempts to avert it.

VI

THE COURAGE OF A PRUDENT MAN

THIS time one hundred and fifty thousand miners came out on strike and only a few thousand remained at work. While the strikers paraded through the streets of the towns with home-made banners, carrying such slogans as "We are slaves now but Mitchell will soon set us free," the man on whose shoulders, whether he liked it or not, the whole responsibility lay, worked in his headquarters at Wilkesbarre and was almost overawed by the magnitude of the task that had been put upon him. In addition to the men made idle in the anthracite field, Tom Lewis had called out twenty thousand in West Virginia, in Alabama there were another twelve thousand out, and in Michigan three thousand miners were on strike,—altogether two hundred thousand men out. It had immediately become so big an affair that in many quarters the temptation arose to make it complete, to call out the whole central competitive field and all the outlying districts. To these people it seemed that the concentrated control which a few financiers had over the country could only be met by a unanimous attack.

Mitchell wanted to see the thing in a large way. He knew that this time the anthracite operators were going to settle once and for all the threat to their power of management which had been raised in 1900. He knew that the hundred thousand dollars in the national treasury would not go far toward feeding two hundred thousand miners and their dependents. This time he had no presidential election to help him. He also knew that the one thing which led doubtful operators to meet him was his insistence that unionization meant stabilization, that contracts were respected. It had long been his trump card.

He felt himself being shoved into a position which he did not like. His chief quality was the courage of a prudent, indignant man rather than that of audacity. He could do nothing, or in the situation preferred to do nothing, to stop the anthracite leaders from asking that the suspension be made national. Only five districts were needed to insure the call of a special convention, and once called, it was doubtful whether such a special convention would dare to refuse the request of striking districts. It would look too much like desertion in a crisis, especially in view of the fact that the national convention had made provision for the possibility of such a suspension and the strike order had spoken of it. Within two weeks after the calling of the anthracite strike, the five necessary districts, West Virginia, Michigan and the three anthracite districts, all of whom were on strike, put a call for the special convention into his hands. Now it was up to him.

Mitchell did not want to use the call immediately. He kept it in his pocket almost a month. He had seen public opinion at work in 1900. He wished to do all he could to avoid alienating it now. To go to the convention now meant that he would face a repudiation of union policy and of himself. Other people saw his own situation. Rumors began to fly that he would resign if a national strike were called. The story of his confidential offer to give Hanna authority to settle for a five per cent increase instead of the twenty per cent demanded was made public. It made many hesitate. Unfriendly papers reminded the strikers that Mitchell had been against their strike, conveying the impression that he was not unwilling that they should lose in order that he might be proved right. At about this time, Hanna announced in Washington that he did not believe there would be a national strike. To all who knew his close connection with Mitchell, this did not sit well.

The situation was touchy. The State troops were in the field now in June, riding hard on the parades. The miners, who had always been vociferously patriotic, were irked at their presence. Lattimer was only five years old, the Molly Maguires still a live tradition in the southern part of the district. For the moment there was order, but it was the tense order of expectancy. Mitchell met the situation by praising the miners for their law-abiding conduct, holding them up to the world, making, as far as he was able, a tradition of peacefulness for them. Their only violence was to be such shock as their patient idleness caused to the susceptibilities of the operators.

What help could be expected from the public opinion on which he counted so much? A group of brilliant writers and sociologists had gathered around Mitchell, many of them volunteering their services. John R. Commons, who had already made a name for himself as the outstanding student of the American labor movement and the first scholar to present it as an inevitable part of industrial development, was sent into the region by the National Civic Federation to study the situation. He brought with him Walter Weyl, later to become collaborator with Mitchell on his *Organized Labor*, his companion on a trip to Europe, and still later to be one of the founders of *The New Republic*. John Graham Brooks, who had been previously interested in the spectacle of oppressions and hopeless hatreds of the anthracite fields, offered his assistance. Somewhat later, Henry D. Lloyd, then the best known humanitarian of the day, using his great wealth in the fight for industrial justice, came to Mitchell with a letter from Father Power of Spring Valley. There were others—newspaper men and women, clergymen, social workers—who offered themselves in the service of the miners' cause. Mitchell had no scorn for these "intellectuals" who showed a willingness to learn from as well as to teach the workers. Among these men and women there grew up a rather astonishing faith in the uneducated, not widely read young leader who, unwilling to wander out into the wider fields of social reform, still embodied the great experiment in that direction. They fed back to him in terms the social worker and humanitarian element of the country could understand the truth he had gained from his own childhood, that poverty and dependence were not qualities that made the men America needed. Who else was there? The press was almost entirely with the strikers. Hearst in New York began to insist that the legal status of the coal-carrying and mine-owning railroads was such as to warrant prosecution under the Sherman Anti-Trust Act. In the White House, Roosevelt called Attorney-General Knox into consultation. Was this another Northern Securities case in which Morgan and the other financiers could be surprised and rudely awakened to the fact that the government might enforce its laws? Could the government start suit? He was informed that the railroads had managed their affairs with both eyes on that law and that suit was difficult. Sentiment against the trusts began to be expressed in an increasing number of newspapers. Hearst pressed his complaint. The man in Wilkesbarre wondered how much all that would help.

There remained President Roosevelt. He let it be known that he had hoped to appoint an arbitration commission but could find no legal authority for it. In June, he sent United States Commissioner of Labor Carroll D. Wright into the region to secure the consent of both parties to an arbitration and to make an investigation. The day before he entered the region a young boy was shot by the coal and iron police. Two weeks later Wright reported that the miners were conciliatory, even willing to forego formal recognition of the union. From George F. Baer he drew a sharp rebuke that the operators had had enough of political interference and threats in 1900 and could get on very well without anything further of the same kind. Wright's own conclusions strengthened the support of the miners. He recommended that a nine-hour day be granted temporarily, with provision for a joint committee of conciliation to make a complete investigation. He also recommended—and the miners had no liking for the suggestion—that a separate organization of anthracite miners be established, independent of the soft-coal fields. The strikers were hoping that the special convention which Mitchell would soon be forced to call would bring the bituminous miners very directly to their support.

The report had no effect on the operators. Now discouraging news came from West Virginia. The suspension had not been complete there. Mother Jones and other organizers had been arrested and were being kept in jail. While they cried to high heaven at the injustice of it all, injunctions and attacks on the strikers remained everyday occurrences there. It began to seem as though the feeling for a sympathetic strike would become inevitable and irresistible. The Illinois district secretary, Ryan, converted a hundred and seventy-five thousand dollars' worth of stocks into ready cash. Delegates representing the four states in the central competitive field met at Springfield. After conferences with his friend, Ryan, and his other friend, Harry Taylor of the Illinois Operators' Association, Mitchell finally issued the call for the special convention to be held on July 17th, a month ahead. He had gained seven weeks from the time the call was placed in his hands, an important gain for a man who hoped to win the strike without going back on a major principle. He had waited long enough for the miracle that did not happen. Now he had to face a clamorous uproar for action.

When he finally stepped before them, the demand for a great, never-to-be-forgotten gesture was still loud, but now men had had time to think it over, to see that the anthracite region could hold out without it and to wonder whether that was the best means of helping the strikers. Some three weeks earlier an unusually competent observer, Walter Wellman of the *Chicago Record-Herald*, had estimated that the ballot on the national strike would immediately line up as many votes from the five states for the strike as there might be in the central competitive field against the strike, and that it was quite possible that a split of votes among the Illinois delegation, due to the Socialist sentiment there, would require but a small number of votes from the scattered regions to bring a majority for the strike. In the intervening weeks, however, most of the doubtful votes had definitely swung against the strike. There were still enough, however, including a percentage of the Pittsburgh district which was on strike, to make the result dependent upon the fight on the floor of the convention. Before it went into session, several State conventions passed resolutions against the national strike.

Mitchell did not wait to allow the convention to become an attack on himself for holding the call so long. He immediately put a plan before them that would avoid "disaster and dishonor," that would avoid destroying confidence, arraying "in open hostility to our cause all forces of society." He appealed to their sense of honor and pride that up until now contracts based solely on the good faith of the union had under the most trying circumstances been kept inviolate. He went further. He took the responsibility of the anthracite strike on his own shoulders. "It will be won without repudiating our solemn agreements."

He looked around over the tense, quiet audience. Not a tall man, lean-faced and worn from the steady grind on his nerves and body, he faced them steadily. Some of his opponents he had satisfied with the promise that the responsibility of the anthracite strike was now his. He knew well enough what capital they would make of it in case that strike failed. Even among the Socialist opponents he had personal friends who did not distrust him or think him weak and dishonest. Their ideals lay in a world shut off from him by his diffidence and his responsibility to the men immediately involved in the strike. He did not want the debate to center too much about the ultimate purposes and aims of the union. He knew that the majority of the strikers wanted financial relief at once. Only if no other alternative was given them would they override the districts which had already let it be known that they wanted to keep their agreements. He gave them that other alternative. His plan went to the heart of the matter as he saw it—finances. The national treasury was to advance fifty thousand dollars immediately to the anthracite districts. Each district was to donate what it could afford from its treasury. Every working miner was to pay a dollar a week. Every officer of the union who received over sixty dollars a month was to make a considerably larger contribution, a quarter of his salary. This was coupled with an appeal to the other unions and to the general public for financial aid and for pressure on the stockholders of the anthracite railroads.

It was a considerable tax that he proposed. It carried its own evidence of good faith. It did more than that. It reminded the delegates that they would have to go home and inform their locals that where Mitchell had said one dollar a week from every miner would for the time being turn the trick of aid, they themselves had insisted that they knew better—with the result that all their neighbors were out of work and unable to support, with money, the strikers in the anthracite fields. Against this proposition they could not afford a great but perhaps futile dramatic gesture of solidarity. The attack of the Socialists, and of those who were already on strike and had nothing more to lose, broke down. All day the discussion went on, Mitchell, following his usual policy of insisting that every delegate who wished it be given the opportunity to speak.

Overnight the central competitive fields caucused and announced themselves ready to vote solidly for Mitchell's plan. The next day some of the anthracite officials came out for it in principle. Now began a flood of contributions by telegram from other internationals, which gave a promise of great results from all the rest of the labor movement. Illinois put fifty thousand dollars into the pot as a voluntary donation, Ohio followed with ten thousand and other states made large pledges. It seemed obvious evidence that Mitchell knew how to do the job in the best way. The plan was accepted and Mitchell got peace with honor. Those who wanted war, open as well as implicit, received a lesson in diplomacy.

While the public applauded the man who held something sacred in a turbulent world, the miners on strike in the field grew high hopes of the relief they were to receive. Some newspaper had estimated that each of the hundred and fifty thousand anthracite miners on strike should receive at least five dollars a week. That meant three and a half million dollars a month. Nothing like this came into the field, but because the story had once been spread the feeling of discontent and suspicion began to spread. Money was getting to be worth too much to trust anyone who might happen to have the handling of it along its road to the strikers' pockets.

The men kept the peace pretty well. No serious attempt to

open the mines had been made. The saloons were less occupied than during normal times. Most of the men had taken a pledge of temperance. Ramshackle houses were being repaired, vegetable gardens dug, baseball games played. Then late in July, in the city of Shenandoah in District No. 9, an overzealous deputy sheriff fired upon a crowd of strikers. He claimed that after parading around the mine stockades they made a rush at the gate. They mobbed him and also a local merchant who came to his rescue. The State troops were already in the field. They marched into Shenandoah and made an armed peace. The miners began to walk with a chip on their shoulders. They were not accustomed to being shot at in this country. The miners had nothing against the soldiers in themselves, for whom the matter took on the aspect of a thrilling picnic, but they did not like the idea of so many guns in the region. National guard, coal and iron police, State troopers—all had guns. Some of the newspapers began shrieking about retaliation and revenge. One of the hostile New York papers hurriedly sent down photographers to get pictures of the incipient revolution. When they failed to see it, they paid men, women and children a little loose change to stand together with hands upraised. Later they were published in New York, and Baer and others seeing them might perhaps permit themselves a personal comparison with St. George.

Mitchell began to lose more sleep. Miss Morris was never allowed to refuse strikers an audience, although their tales of suffering helped to unnerve him. The thing had gone on longer than people thought possible. In 1900 the skirmish had been won in six weeks. Now the third month was dragging out. He began to feel a personal responsibility for every bit of suffering brought to him. In addition to the insomnia there was also pain from an old sickness. The waiting was telling on him as it was on all the others. Now again stories began to be circulated. The delay and the magic five dollars that was to come to every man had its part in starting them. Nasty stories of misappropriation of funds circulated, old women's gossip that he was hardly the saint he was painted, that at a time when many of the miners had taken the pledge, Mitchell was drinking. Credible details were added. On top of all this, one of the priests, a Father H. F. O'Reilly, wrote to one of the newspapers: "Everyone was happy until Mitchell and Fahy came and organized the union; these men are not workingmen; they are not respectable, and I would not give two cents for the honor of either of them. They draw big salaries and live on the best in the land." Another priest, Father Power of Spring Valley, Mitchell's old adviser, rushed to his defense. There was correspondence back and forth, all eagerly read, serving for want of better news. The priest finally admitted that he had not meant to reflect on Mitchell personally. The reporter of an unfriendly newspaper included the five-course menu of the modest hotel at which Mitchell was staying, a story which the copy-reader headlined for the benefit of the strikers, "What did you eat to-day?" The reporter was hooted. So many attacks began to arouse loyalty that had grown self-centered in the days when every man had to think first of his own troubles.

Mitchell knew that something which might be tne breaking point had come now early in August. A year later he wrote: "I am fully convinced that the strike would have collapsed had the operators at this time opened their mines and invited the strikers to return." Then financial aid began to pour in as a result of the action of the special convention. Sums were hurried to the weaker points. Circular letters were sent out explaining the situation. Men were encouraged to find work elsewhere and left on "Johnny Mitchell" specials.

At this time George F. Baer came to his rescue. One of the local citizens had appealed in the chastened, almost solemn mood of the strike that Baer should do what he could to end it. Baer took the occasion to set him right on the essential issues involved. The letter was, after some time, given to the press, which immediately broke out into vehement editorials, cartoons and exclamation marks. It seemed to have included so completely what everyone suspected was in the operators' minds, to show so fully the gap between those who were being called the common people and those who were being called the coal barons that it was recognized at once as a social document. It is said that Baer denied the letter in the manner later made popular by some politicians toward reporters who quoted some slip of theirs. It did not work. The letter had been photographed. It read:

"Philadelphia and Reading Railway Company
"President's Office,
"Reading Terminal, Philadelphia.

"17th July, 1902.

"My dear Mr. Clark: I have your letter of the 16th inst. I do not know who you are. I see that you are a religious man; but you are evidently biased in favor of the right of the workingman to control a business in which he has no other interest than to secure fair wages for the work he does.

"I beg of you not to be discouraged. The rights and interests of the laboring man will be protected and cared for—not by the labor agitators, but by the Christian men to whom God has given control of the property rights of the country, and upon the successful management of which so much depends. Do not be discouraged. Pray earnestly that right may triumph, always remembering that the Lord God Omnipotent still reigns, and that His reign is one of law and order and not of violence and crime.

"Very truly yours,

"George F. Baer,

"*President.*"

That God was on the side of the great industrialists of the country was something the country at large was not yet ready to admit. Great industrialists might have the money of the country, but the common people still had God. Just a year before, when J. P. Morgan had expressed some satisfaction with the country, Bryan's *Commoner* had taken occasion to comment that if the time ever came when he was dissatisfied with it, he might return it to the people. Now parallel columns drew a comparison between Baer's statement, "George the Last" as Darrow was to dub him, and the words of George III.

The financiers might be ridiculed, but they held control. Mitchell himself did not comment on the Baer letter until much later and then mildly. The operators began to use a few strike breakers, housing them on the mine property to protect them from the workers. Here and there a few men who had been on strike gave in and went to work, stealing from their houses at night. A few households attempted to move from the towns onto the mine property. Their wagons were upset in the middle of the street. Strikers refused to go to the same church with them. A few outhouses were dynamited—a grim sort of humor. Every so often someone would lose patience with the troops protecting the strike breakers and throw a ripe tomato their way. General Gobin ordered the troops to shoot without further orders any persons who harassed them by throwing missiles. Mitchell again encouraged the miners to keep the order.

At the very moment when Mitchell was despairing of the success of the strike, there came a test as to the terms on which he might encourage the workers to go back to work. Ralph M. Easley,[1] secretary of the National Civic Federation (which had on its board Hanna, Gompers, Oscar Strauss, as well as others more conservative on the labor question) renewed efforts to find some formula for compromise. He found that the railroad presidents had their backs up, that they "were willing to put in millions for defense, but none for tribute." George W. Perkins, a partner in the Morgan firm and a friend of Roosevelt's, made a suggestion which Easley thought acceptable. It was that Mitchell send the men back to work, while a committee appointed by Hanna—or a committee with Hanna, Frank P. Sargent, former labor leader and now Commissioner of Immigration, and Charles M. Schwab (or someone else representing Mr. Morgan)—make an investigation, with the aid of a financial man, a mining engineer and some sociological expert. This committee would then wait upon Mr. Morgan, who had said at one time that he would "do what was right" to urge his acceptance of their conclusions. Mitchell was to have a chance to approve or dis-

[1] Letter to Frank P. Sargent, August 4, 1902. Courtesy of American Federation of Labor.

approve of the personnel. The men who drew this plan counted too much on Mitchell's tiredness, on the outburst at Shenandoah, on the belief that the entrance of thousands of the State troops had lessened the chances for keeping up the strike. Mitchell was a conservative, but he would not endorse any plan to put the strikers back to work on the hope of an unofficial committee of such a type which would in turn have to count on Mr. Morgan to accept their conclusion. Mr. Easley's proposal was never made public.

People in the city of New York and elsewhere were beginning to feel the strike in another way. Coal that had been laid down in New York at five dollars a ton was climbing to fourteen, and the poor who bought it in small lots were paying for it at the rate of twenty dollars, and even that supply threatened to give out. As September approached, they read the papers with an again renewed interest.

The merchants in the anthracite district had long since passed the limit of extending credit of their own to their striking customers and were now finding the banks unwilling to carry them any further. They petitioned President Roosevelt to intervene in some way, in any way, but to intervene at once. There were now continual parades of soldiers through the mine towns, greeted by cries of derision. Union leaders went from door to door to warn the strikers to stand the insult of being policed in this fashion. A general order was issued allowing the soldiers to arrest girls who insulted them. That turning of the tables brought a momentary laugh. Some priests in the region initiated a movement to have the troops withdrawn. The miners had the satisfaction of knowing that all the bayonets in the world could not hew a ton of coal.

Politicians began to stir again. It was not a presidential year but Senators Quay and Penrose of Pennsylvania, Senator Lodge of Massachusetts did not want to lose hold in the forthcoming state elections. They began to feel the pressure from below as the fall drew closer. Senator Platt of New York and Mayor Low of New York City showed great concern over the situation. Prices continued to soar to twenty dollars a ton and, for those who bought it by the bushel, to unheard-of amounts. Governor Stone of Pennsylvania was expected to call a special session of the legislature to consider a compulsory arbitration bill. Nothing came of that. There was a brief moment of hope when Senators Platt and Quay made a statement that the strike would end in two weeks. Wall Street took it seriously enough to finance a small boom in anthracite stock.

Every day brought some often good-hearted and usually futile attempt to influence the situation. The Socialist Party held special conferences; Jay Cooke suggested that a ten per cent increase and a nine-hour day would be satisfactory to the operators if they would be allowed to increase the price of coal twenty cents. In Chicago *turnvereins* and religious conferences aided and prayed for the miners. The New York Democratic State Convention went on record for the nationalization of the mines.

At this moment the West Virginia strike in the Kanawaha and New River districts, under the leadership of Tom Lewis, came to an end. Each local was advised to settle on the best terms it could get. The strike had lasted almost as long as the anthracite struggle to date. Its disastrous end was an open advertisement to the anthracite operators that they too might be able to push people to the limit of human endurance. The bad news followed a few days after a statement by Baer that "We are fighting the battle of freedom for the individual and the right to labor on his own terms." He would resign rather than arbitrate. John Markle, who in 1900 had urged his miners to arbitrate, was now evicting men from his houses. He said: "There will be no arbitration. Mitchell has become a demigod and alone is obeyed, so that all ordinary and usual matters of discipline are practically disregarded." It was a firm stand from both the railroad companies and the independents.

But now the miners knew that they were making history, and they wanted it to be a proud history. The end was further in the distance than they saw, but now they knew they would stand together until the end. It was at this moment, after four months of strike, that Wellman reported that there was not a breath of criticism. If Mitchell, he said, "were to tell the men to go jump in the Wyoming River they would do so. They have no more doubt that he is going to win the battle for them than they have of their being in America. . . . To a great many of the newly arrived miners John Mitchell is the one great man in the United States. Possibly they have heard of Pierpont Morgan and they have a dim idea there is such a man as President Roosevelt, but ask the first Hun or Polander on the streets who is president of the United States and the odds are about even that he will reply 'Johnny D'Mitch.' "

Now Mitchell could announce that the miners had enough money to fight it out on these lines all winter. The bituminous districts had donated a quarter of a million and had sent in assessed contributions of two millions; $420,000 had come in from other trade unions. Mitchell's own district, Illinois, had sent in in donations $861,000, almost twice as much as the outside unions and the general public put together. In January, 1903, there was still over a million dollars on hand in the treasury.

It was getting toward the end of September. On the 26th of the month the Philadelphia and Reading invited its employees to return to work or lose their jobs. On September 28th Governor Stone announced that he would use all the State troops to protect men who wanted to go back to work. The next day Mitchell and the three district presidents were conferring with their lawyers on the possibility, so it was said, of action for government ownership and operation. There were still over two months to go until Congress convened. Already sentiment for a special session was growing. That day it was announced that President Roosevelt was taking the matter in hand.

Theodore Roosevelt had come to the presidency from that oblivion into which Senator Platt, boss of New York, had tried to send him. The bullet that took McKinley away put into power a man who had not been particularly identified with the aims of the working class and who had condemned many of their leaders. He spoke the language of and appealed to the social reformers, and that "middle class" which felt itself losing out in the new world so completely under the control of the financiers and industrialists. He combined this reputation with one of a man of action—founder of the club of men who first went up San Juan Hill—a hero on fields where there was not much doubt as to who was victor and who vanquished. He had used a great deal of vigorous language about iniquitous wealth, but he depended on publicity mainly to frighten the financial powers who were so largely at that time the supporters of the Republican Party. He saw himself limited in the face of the suffering of the strike in many ways which irked him. His attorney-general had told him that the coal-carrying railroads had managed their combination with the mine properties so skillfully that he could not threaten them as he had Morgan in the Northern Securities case that February. On the other hand, if, as he wrote Lodge, he was "more genuinely independent of the money interests in all matters where I think the interests of the public are concerned, and probably I am the first president of recent times of whom this could be truthfully said," he recognized that he had earned no favors for which he could now ask.

The use of troops in which he placed great confidence could not be invoked without the request of Governor Stone, and the Governor, suspecting the use to which Federal troops might be put, made no such request. The question of whether Congress had a right to legislate concerning an entirely intra-state industry such as the anthracite was so debatable as to preclude much hope from a special session of Congress. Meanwhile he knew, and wrote Hanna, whom he had publicly berated in 1897 for hesitancy about the country's entry into war but with whom he was now friendly, that the public at large would "tend to visit on our heads responsibility for the shortage in coal, precisely as Kansas and Nebraska visited upon our heads the failure to raise crops in the arid belt eight, ten or a dozen years ago." The governors of Massachusetts and New York notified him that as the cold weather came on, if the coal famine continued, the misery throughout the northeast and especially in the great cities would become appalling and the consequent public disorder so great that frightful consequences might follow. He commented at this point in his autobiography: "As long as I could avoid intervening, I did so. . . ." He wrote his friend Hanna to please bestir himself again in New York. On September 27th Hanna replied: "Confidentially I saw Mr. Morgan and I also saw Mr. Mitchell (the public knows nothing about that). I got from Mr. Morgan a proposition as to what he would do in the matter and I got Mitchell to agree to accept it—if the operators would abide by the decision. I really felt encouraged to think I was about to accomplish a settlement." He found, however, "to my surprise," that Mr. Baer absolutely refused to "entertain it." Hanna felt that the operators would stick it out until the "miners are starved to it, and that may be weeks ahead as they are getting abundant supplies from their fellow workmen all over the country." Hanna felt that the operators had "put all efforts of mine in a false light before the public." With the knowledge that Morgan and the railroad presidents themselves were no longer one in the situation, Roosevelt decided to use the only method open to him, the unofficial one of moral suasion. On September 29th he sent to Mitchell a telegram stating that he would "greatly like" to see him in Washington on October 3rd, "in regard to the failure of the coal supply which has become a matter of vital concern to the whole nation." The presidents of the coal companies received similar dispatches.

On Friday, October 3, 1902, at eleven o'clock in the morning, Mitchell (who came with the three district presidents) and the operators walked through the large crowds waiting outside the doors of the temporary White House. Roosevelt, still confined to his chair from injuries received in Pittsfield, Massachusetts, where his carriage had been struck by a trolley, received them. Mitchell, about whom he, together with the rest of the American public, had read for so long, was haggard and seemed painfully young in that gathering of impressive middle-aged and elderly officials and capitalists. Roosevelt scrutinized him closely. Here too was a fight leader, without the advantage of military discipline to help him, a man whose personal power led others on. Roosevelt had tasted a little of the trust Mitchell's men gave to him and he knew how precious it was and how hard to earn. He had read the solemn words Mitchell had spoken in Philadelphia the day before he had been summoned to Washington: "We are conducting this struggle without malice and without bitterness. . . . We ask not for favors but for justice." Then came language which Roosevelt, who said he stood not for the poor man nor for the rich man, but for the righteous man, would never use:

"Involved in this fight are questions weightier than any question of dollars and cents. The present miner has had his day. He has been oppressed and ground down; but there is another generation coming up, a generation of little children prematurely doomed to the whirl of the mill and the noise and the blackness of the breaker. It is for these children that we are fighting. We have not underestimated the strength of our opponents; we have not overestimated our own power of resistance. Accustomed always to live upon a little, a little less is no unendurable hardship. It was with a quaking of hearts that we asked for our last pay envelopes. But in the grimy, bruised hand of the miner was the little white hand of the child, a child like the children of the rich, and in the heart of the miners was the soul-rooted determination to starve to the last crust of bread and fight out the long, dreary battle to the end, in order to win a life for the child and secure for it a place in the world in keeping with advancing civilization."

With Mitchell were the three district presidents, T. D. Nicholls, commonly spoken of as a radical and accounted the most thoughtful of the three; Thomas Duffy, short, vigorous, representing the older rough-and-ready element; and John Fahy, the handsome orator who had once been nominated for the vice-presidency against Mitchell. They brought with them the hopes of a hundred and fifty thousand men.

For the operators there appeared George F. Baer, whose letter had made him more famous than his position in the industry ever could have done; W. H. Truesdale of the Delaware, Lackawanna and Western, who was said to have been playing golf the day the strike was declared; T. P. Fowler of the New York, Ontario and Western; David Wilcox of the Delaware and Hudson Company, who was later to use his legal training to grill Mitchell unmercifully; and John Markle, the independent operator, as a representative of his group. For the government, there was Attorney-General Philander

Knox, Secretary of Treasury Cortelyou, who was later to become president of one of the largest corporations in America; and Carroll D. Wright.

After the usual statement of the seriousness of the situation and thanks for the presence of the representatives, Roosevelt said that he realized he had no legal right or duty to interfere but that the urgency of the situation impelled him to use whatever influence he could. He stated that he did not invite a discussion of the claims. "I appeal to your patriotism and to the spirit that sinks personal consideration and makes individual sacrifices for the general good." Mitchell answered first for the miners. He too was impressed with the gravity of the situation, he said, although the miners were not responsible for it. The miners were willing to meet with the operators to adjust differences. If that could not be done, "we are willing that you shall name a tribunal which shall determine the issues that have resulted in the strike and if the gentlemen representing the operators will accept the award or decision . . . the miners will willingly accept it even if it should be against their claims." This was to Roosevelt's own heart. He wanted ways of action discussed. The operators wished time to put their answer in writing and there was a temporary adjournment.

Mr. Baer read the answer of the coal operators when the conference reopened. He did not understand how the discussion could proceed along the lines indicated by Mr. Roosevelt, "but we assume that a statement of what is going on in the coal regions will not be irrelative." He then proceeded to picture a "terrible reign of terror, lawlessness and crime," from which "only the lives and property of the members of the secret, oath-bound order . . . are safe." He spoke of the dynamiting bridges, maltreatment of the fifteen to twenty thousand men he said were at work by those forces over whom "John Mitchell, whom you have invited us to meet, is chief." He outlined the duty of the president to him. It was "not to waste time negotiating with the fomenters of this anarchy and insolent defense of law, but to do as was done in the war of the rebellion—restore the majesty of the law.

"The government is a contemptible failure if it can only protect the lives and property and secure the comfort of the people by compromising with the violators of the law and the instigators of violence and crime. Just now it is more important to teach ignorant men, dwelling among us, misled and used as tools by citizens of other states . . . that order must be restored. . . ." He declined Mr. Mitchell's "considerate offer to let us work on the terms he names." He offered to continue the wages existing at the time of the strike and the adjustment of the wages at each colliery, and in addition, to submit disputes to the judges of the Court of Common Pleas of the district in which the colliery is situated for final determination."

Now it was Mitchell's turn again. He wanted to be very definite on one matter—the miners were not suggesting arbitration because they were beaten, because they had any doubts of the justice of their claims, but in the hope of relieving the situation for the consuming public. He was confident that any impartial tribunal would see the equity of the miners' demands, but whether or not it did, its decision would be accepted, if it were also accepted by the operators in the form of an agreement for not less than one nor more than five years. He would then call a convention of the miners and recommend the return to work, with the understanding that the wages decided upon were to go into effect from the day on which work was resumed.

Markle arose for the independent operators. He repeated Baer's stricture and concluded: "I now ask you to perform the duties vested in you as President of the United States, to at once squelch the anarchistic condition of affairs. . . . Are you asking us to deal with a set of outlaws? I can hardly conceive of such a thought." The operators leaped into the rôle of defenders of the faith with considerable ardor. To their charge of twenty murders, Mitchell answered that if it could be proven that the strikers had committed such alleged murders, he would resign his position. He admitted some lawlessness but said it had been provoked by the criminal element composing the coal and iron police.

The operators did not accept his offer. They stood by their guns. They would refuse to recognize him in any capacity whatever in the settlement of the strike.

Then everybody went home. Roosevelt wrote Hanna that he was downhearted over the result. It was a blow to his own prestige. He said later: "There was only one man in that conference who behaved like a gentleman and that man was not I." He who was so articulately impatient was impressed by the patient silence of Mitchell in so desperate a situation. He was discouraged both because of the "great misery ensuing for the mass of our people and because the attitude of the operators will beyond a doubt double the burden on us, who stand between them and socialist action. . . ." He began to consider what to recommend to Congress and added, in his letter to Hanna, that the attitude of the operators made him wonder if he could not make a "fairly radical experiment" on the anthracite coal business to institute some power of supervision and regulation by the government over such corporations. He told Hanna that had the operators acceded to his plan, he would have put him on the commission. Later he wrote ex-President Cleveland that he had him in mind as one of the members of such a board, in fact, as its head.

Conferences began all over again in New York and Philadelphia. Mitchell went back to Wilkesbarre. Roosevelt called on Governor Stone to act under the State law against the railroads. The mine operators visited Secretary Root in the hope, it was rumored, of getting Federal troops. Oscar Strauss tried to solve the situation by suggesting that Roosevelt accept the common pleas judges Baer had mentioned as arbitrators and wired Mitchell urgently to accept the proposition. Mitchell replied that an arrangement for three hundred and sixty-four separate arbitrations for each colliery was out of the question. Governor Stone kept on sending more State troops into the region so that the "men could go back to work." None went.

Roosevelt, piqued, refused to be "turned down by the operators" and talked of "little else than the coal strike and virtually dropped all other public business for the time being." If Morgan's interest was not enough, he would go after President Cassatt of the Pennsylvania Railroad. He did. The strike continued. He got a little comfort from a laudatory letter from Grover Cleveland, who, as President, had sent Federal troops into Illinois over the protest of Governor Altgeld and had broken the back of the Pullman strike. Roosevelt's appreciative reply was that on going over Mr. Cleveland's action he decided to "try to use this force with the same firmness that you showed. . . ." Meanwhile, since he had no success with Governor Stone or the operators, he tried Mitchell again. On October 6th he sent Carroll D. Wright to propose that "if Mr. Mitchell will secure the immediate return to work of the miners in the anthracite region, the President will at once appoint a commission to investigate thoroughly into all the matters at issue . . . and will do all within his power to obtain a settlement of these questions in accordance with the report of the commission."

Mitchell, for all his willingness to be conciliatory, would not accept the proposition. He had seen what little effect Wright's recommendations had made months ago. Nothing new had developed in the Washington conference to suppose that the operators would accept a report after it had been made. Hanna wired Gompers to use his influence over Mitchell, to which Gompers replied that he could not "advise Mr. Mitchell to accept a proposition so elusive and unpromising." Hanna could not dictate to either side.

Then Mitchell brought the men together to test and snow their solidarity. Their convention voted unanimously in favor of continuing the strike "until the matters in dispute between them and their employers were decided by a board of arbitration, even if all the troops of the United States were sent into the anthracite coal fields." The men stood solid but would accept arbitration.

As a matter of fact, Roosevelt, according to his autobiography was at the moment planning to send "all the troops of the United States into the anthracite coal fields" over the protest of the miners. He described his plan: The Governor of Pennsylvania would "ask me to keep order." He would then "put in the army under the command of some first-rate general." He had the man all picked out, a benevolent person, "a most respectable-looking old boy," Major General Schofield. No interference by the strikers with men who wanted to work was to be permitted. The operators would be dispossessed and General Schofield would run the mines as receiver "until such time as the Commission might make its report, and until I, as President, might issue further orders in view of this report. . . . At that time I should have been almost unanimously supported. . . ." Governor Stone, however, gave him no opportunity to try this army game.

Meanwhile the country was talking and praying. The day after Mitchell refused the President's second proposition, he went to New York, by request, to see Senators Quay, Penrose and Platt, Governor Odell and other politicians. Now Cassatt of the Pennsylvania was to be influenced through Mayor Low by receiving a much-desired franchise for the Long Island Railroad. But Baer told Governor Odell of New York again, "We will not accept political advice or allow the interference of politicians."

More drastic action was now discussed. In Detroit a conference of mayors and governors called to consider the situation actually passed resolutions in favor of a receivership to be declared by Congress at a special session, and plans were outlined for the raising of funds to aid the strikes. The Western Federation of Miners wired Mitchell their solution that all the miners in the country be ordered to strike. The American Federation of Labor sent out a widespread appeal for money, which contained a clause remarking that the miners had been willing to leave the entire matter to Mr. Morgan and "what more could the miners do and maintain their self-respect?" Mitchell requested that this be deleted and it was.

It was still a month and a half before the convening of Congress. Public sentiment had been so vocal that there was very little doubt that some plan which would take away the absolute control of the industry from the operators would be proposed and passed and put to Roosevelt, who had already said so much that he could not be counted upon to oppose such Socialism. Secretary Root asked permission to mediate in his own name and had a five-hour conference with Morgan on his yacht, while Baer and Cassatt conferred elsewhere. The result was that on October 13th, ten days after Mitchell had been received at the White House, Mr. Morgan and Robert Bacon, a member of his firm, entered the White House and announced that the operators were at last willing to submit to arbitration.

The country was one loud clamor of approval. It was about time. In Chicago the people had been ripping up the wooden paving blocks for fuel.

At Washington it became a question of the composition of

the arbitration commission. Mitchell insisted that he could not relinquish, for the protection of the miners, the right to have one representative of labor on it. The idea of having the board made up mainly of nominees of the employers was already drawing fire from the mine fields, where the men called local meetings and denounced the scheme as a gold brick. Roosevelt saw the point but he had his own troubles with the operators who had "worked themselves into a frame of mind where they were prepared to sacrifice everything and see civil war in the country rather than back down and acquiesce." In his autobiography he wrote of the deadlock: "Suddenly after two hours' argument it dawned on me that they were not objecting to the thing but the name. . . . I shall never forget the mixture of relief and amusement I felt when I thoroughly grasped the fact that they would heroically submit rather than have Tweedledum, yet if I could call it Tweedledee they would accept it with rapture; it gave me an illuminating glimpse into one corner of the mighty brains of these 'captains of industry.' . . . All that was necessary for me to do was to commit a technical and nominal absurdity with a solemn face. This I gladly did. I announced at once that I accepted the terms laid down. With this understanding I appointed the labor man I had all along had in view, Mr. E. E. Clark, the head of the Brotherhood of Railway Conductors, calling him an 'eminent sociologist,' a term which I doubt he had ever previously heard. . . . In publishing the list of the Commission when I came to Clark's appointment, I added: 'As a Sociologist, the President assuming that for the purposes of such a commission the term sociologist means a man who has thought and studied deeply on social questions and has practically applied his knowledge." Roosevelt also added Bishop Spalding of Peoria, and, as recorder, the person whom the operators had expected him to name as an "eminent sociologist," Carroll D. Wright.

The operators had accepted, in their fashion, and now Mitchell, despite the fact that the previous convention had directly endorsed arbitration, took no authority on himself. He sent out a call for another convention of the anthracite delegates of all three districts, but he also issued, in the name of the Wilkesbarre district, the acceptance of that district. Immediately he was deluged with criticism. Many of the local leaders felt that the submission of the operators proved conclusively that they were whipped, and since the miners had the funds, it was folly to submit to arbitration with victory at hand. Mother Jones burst into a stormy denunciation of Mitchell. From England came word that the president of the British Miners' Federation, Benjamin Pickard, Member of Parliament, gave it as his opinion that Mitchell had erred gravely.

Mitchell felt that he could have taken no other step without breaking his word and the word of the miners who for months had been insisting on their willingness to arbitrate. To withdraw now because victory was in sight might bring some immediate gain, but it would alienate that very public sympathy which had in his opinion helped the strike. He took silently the taunts that Roosevelt's reception of him in Washington had turned his head. The judgment of most students of the labor movement, years later, was that despite serious defects in the final award of the commission, submission to arbitration had been one of the wisest moves Mitchell ever made, not only for the miners but for the labor movement.

And now the strikers met, five months and a week after they had gone out on strike, on October 20th. Was this the victory they had hoped Mitchell would bring? The debate was long and hot. It lasted for two days. What guarantee, delegates asked, would there be of the payment of back wages, of freedom from discrimination? Would all men get back their jobs? Mitchell answered the engineers and pumpmen—who had not come out originally with the miners but had followed them a few days later when their own demands were refused—by telling them that their rights would be respected. He felt that men who voted should be aware of the fact that a few of them would not be "restored to their former positions." According to reporters, the debate was interrupted again and again by appeals to him to close it, but he always refused, until every man who wanted the floor was heard.

When they had finished, he addressed them. He was weary and looked broken in health. They had suffered much and he praised them for their noble work. He too had suffered. He spoke of his gratitude to the other workers who had helped, and to the many who, though they lacked fuel, did not turn against the strikers. He emphasized particularly the fact that he would himself have preferred direct conferences with the employers, except for the promise of the miners and the earnest pleading of President Roosevelt. He admitted the many objectionable features in the plan as it stood. He said he would not have accepted it at all if labor had not been given a representative to the commission.

Then to that vast multitude of suffering men, he spoke—to the disgust of radicals and to the amazement of those of his people who were and well could be bitter—of his "lack of malice" for their opponents, and of his belief that "there should be no irreconcilable conflict between capital and labor." He had been called a Lincoln, and, as a matter of fact, he had as little of the revolutionist in him as had Lincoln, who had so often disappointed the abolitionists.

At the end of two days the convention accepted the appointment of the commission, and adjourned after the singing of the national anthem.

Mitchell knew how far from complete the victory really was, and raised no hopes that would be boomeranged back to him later. There was no recognition of the union as yet, but there was the open action of the President of the United States to assure the miners that they would have justice. Their story had been greeted throughout the whole five months of strike by the public in a way to convince them that no one who heard their story would fail to see that they needed a new deal. They had been vindicated in the faith they gave their leader, and he had been vindicated in his course of action. He took occasion after their decision to tell the miners that the "strike itself has demonstrated the power and dignity of labor," and perhaps as an answer to the radicals in and outside of the union, "conservative, intelligent trade unionism has received an impetus, the effects of which cannot be measured." To the country at large he pointed out, by way of a letter to the President, that, "if our offer of arbitration had been accepted six months ago instead of now, there need have been no strike." But he saw that something more than a vindication of the reasonableness of their claims had taken place. In this letter he added: "The poor underpaid workers of these coal regions . . . have taught these corporation managers a lesson in civic and social duty." Nevertheless he bore them no grudge for their arrogant accusations and insults.

There was credit enough for all. President Roosevelt's admirers gave him all the glory. But the foreign-born miners still knew only one president, the one who had led the strike and who now presented this settlement. When Mitchell left Wilkesbarre for Washington he was "besieged by hundreds of mine workers and their friends who almost pulled him to pieces . . . and literally carried him off his feet." The Slavish, Polish and Lithuanian groups united in presenting him with a watch; the three districts gave him a diamond ring. The speech of the foreign miners began with the phrase, "Blessed be the moment . . . when you arrived in our midst." A writer in *The Outlook* went so far as to say that he heard in the streets of Scranton a conversation between two Slav miners in which one of them disputed, "Mitchell am more than God."

Throughout the country, too, the praise of Mitchell was overwhelming. A boom was begun to have him elected as President of the American Federation of Labor at its November convention. He was spoken of for a half dozen political appointments and nominations. Newspapers and periodicals vied with each other for terms of superlative commendation. All seemed agreed that Mitchell "had gone through the long struggle with not one error. . . . Without doubt John Mitchell is the greatest labor leader the country ever saw . . . great because he is honest, able, sincere, conservative." There was dissent from this judgment still among the radicals in his own organization and outside of it, who were outspoken in their belief that for one cause or another, he had fumbled in carrying through the opportunity for a great revolutionary strike.

The strike was over. On October 23, 1902, the miners answered the steam whistles after one hundred sixty-three days of strike. Estimates of the cost to the mines and railroads were put at one hundred million dollars. No one could estimate the cost to the strikers themselves.

Now it was over, and for a few moments at least, the man who had carried the final responsibility on his shoulders breathed easily. But there was little rest. As the greatest strike in American labor history was ended, Mitchell began the work of preparing the material for the miners' cause, so that as their advocate he might make a showing worthy of it and of their courage.

VII

BRINGING FEUDALISM TO TRIAL

NOT only by the miners but by the public the appointment of this commission was held to be a victory. While the miners went back to work and to the risk of accident, the members of the commission were blooded by going down into the pits, "descents into infernos," as one of them expressed it, and the sociologists whom Mitchell had grouped around him were preparing evidence on which the industrial feudalism could be indicted and convicted in the open court, Mitchell was being deluged by a chorus of acclaim and appreciation from all over the country.

It would have been enough to turn the head of a weaker man. Louis D. Brandeis, John R. Commons, as well as Henry D. Lloyd, saw in his face the spirit of a man inspired by a great cause. One of the men who wrote most extensively about him was Walter Weyl. Weyl saw in Mitchell more than an instinctive honesty, sincerity and sympathy. He credited him with keenness of perception and saneness of judgment, as well as with a remarkable gift of being able to arouse enthusiasm in others without being blinded or misled himself by that excitement. His very limitation—his single-minded absorption in unionism—was a source of strength. He saw him as a man too sensitive about personal attack on himself to accept such opportunities as his opponents gave him to inflict wounds of that character. "His philosophy," Weyl says, "is like that of the trained prize fighter who feels no hatred against his antagonist, even though he is endeavoring to disable him temporarily." One of the many incidents which influenced Weyl's enthusiasm for this suddenly great figure on the American scene was an occurrence about which he wrote only later.

Apparently there had been people not yet convinced that the old order had changed at all in this man. Their representative, with a large sum of money, went to Mitchell's hotel to induce him not to interfere with the naming of the commission. Weyl described him as "bold, facile, plausible, with a record which might have been envied by the most enterprising of scoundrels." After a short fifteen minutes, he left the room without having mustered up courage to propose his dishonorable scheme to the man who gravely and innocently conversed with him.

One newspaper correspondent pronounced him "not a promising object of study. It is too much like studying a well." To the public at large he was a romantic figure. To them his rise from complete misery to his calm and patient self-confidence at the White House had an enviable and dazzling speed to it, almost as great as if he, like the proverbial poor boy, could rise to the presidency himself. He was frequently described as very handsome, full faced, clean shaven, with deep-set, luminous eyes, firm mouth and high forehead, with brown, almost black hair brushed carelessly back by his fingers. His frock coat and a black tie almost large enough to hide the white of his shirt-front gave him the semblance of a priest.

Even to those who knew he was no saint, there was something of the saint in his character.

Pictures taken during that time do not belie them. In sharp contrast to the somewhat militant aggressiveness and sociability of other labor leaders who had at one time or another taken the public eye (a type much later to be called "extravert") his face revealed a curious combination of hopefulness and strength. The eyes that looked out from his almost swarthy face were not yet the eyes of the "prudent business man" he liked to consider himself, nor yet were they the eyes of the fiery evangelist, who loved humanity well enough to be able to forget, for the moment of speaking, its weakness and his own, and who aroused, awakened and stirred great crowds. Not only Debs, but Gompers, who had not yet become the publicly recognized executive of a going concern, had more dramatic bearing than he.

In a way there was something brittle about him even now, when he was only thirty-two years old. He had worked so hard, life had had so few compensations, that the great quality of humor, which has made life bearable to men greater than himself and has prevented others from taking any cause, even that of the disinherited, with the single-minded solemnity he gave to it, was lacking. He was risking himself into worlds where few men with his meager education ever arrived, challenging and meeting the great powers of the world, and it brought to him an awareness of his own limitations, a persistent mistrust of himself, which might have turned into bluster but in his case had turned into modesty. The constantly reiterated need of his people for his help carried him along. It was a responsibility, directly and intimately placed on his shoulders by each of the thousands of miners he met, a burden he could not shift to another. It was no laughing matter.

In spite of the somewhat priestlike garb and mannerisms, he was not religious in the commonly accepted sense of the word. In that respect he differed somewhat from his people. Welsh miners and Catholic immigrants alike, they had a God to whom they could take their hopes in prayer, and a church where their doubts could be dissolved. Many of the Welsh and English-born mine leaders were preachers in addition to being miners, and the meetings over which they presided had a decidedly religious tone. Most of his best friends—Ryan and Father Power, as well as Mitchell's own wife—were both Catholic and religious. During those days whatever faith he possessed was in the belief that oppressed men could rise to regain freedom and a good life through the agency of the union.

The life he was leading was not entirely that which he had hoped for when, as a young and lonely miner, a man who had boarded in this house and that, he came back to Illinois and married Catherine O'Rourke. Now he was almost constantly separated from her and the four children who were still alive. He saw them only two or three times during the year. He knew how short the life of a leader in the movement could be, and the idea of uprooting the home, severing all the ties of friendship for life in Indianapolis, where he was only present occasionally, hardly seemed worthwhile. There were whispered rumors of the same type which surround every leader who has opposition, largely to the effect that the absence of his wife in the anthracite fields indicated certain deviations from the desired marital code. He avoided carefully what seemed to be attempts to frame him with women. The only women he knew were those with whom he could share his passion for the union. He had no small talk, no ease for others. He had the sentimental American attitude toward wives and mothers—they were on a pedestal.

Much more persistent was a rumor that he was often intoxicated, especially during the strike. Associates who have been frank about his habits at other times have vigorously denied that he drank at all during the strike. During some of these days he suffered from insomnia and its exhaustive effects.

What amusement he did get was not often from books. Intelligent, he was not intellectual nor a constant reader at this time. Books like *David Harum* and Mark Twain's *Joan of Arc* might interest him, but in the main he who could memorize the terms of a mining contract on first reading, had to put a mark on the chapter so that he would not read it again. He enjoyed the theater, but in Scranton and the other cities where the business of the union was carried on, this source of pleasure was limited.

He did like poker, one of the gambles of life where nothing but money is at stake, and was joined in this by a group of men outside the labor movement. Round-faced, jolly John Loftus of Scranton, whom in later years Mitchell often visited, was one of them. Walt McDougall, then beginning his career as a cartoonist, was another. Frank Ward O'Malley, who was later to become editor of *The Sun*, the daily then most unfriendly to both Mitchell and Roosevelt in New York; Dan Hart, actor and later mayor of Wilkesbarre, who twenty years afterwards was to extend the freedom of his city to miners after yet another anthracite strike, when they were still fighting for complete and absolute recognition of the union—these were all men with a large capacity for enjoyment that Mitchell could appreciate and envy.

There was not much relaxation possible before the commission opened its hearings. The miners were not by any means satisfied with the personnel of the commission. The operators were employing an imposing array of legal talent. The public expected something more than a huge theater piece. In many quarters there was high hope that some word would be said, some clear call be uttered that would do more than challenge the triumphant industrial feudalism of the day. Their hope was that here and now some banner could be raised and carried out to the steel mills, the railroad junctions, the textile towns, the sweatshops of New York, the lumber camps of Michigan, and be turned into bread which the oppressed workers could really eat.

The public had already judged the case, but the mechanism of having its representatives bring some constructive effort to the judgment was now to be gone through. The miners had known their case so well that they had been willing to spend every cent they had to fight those whom they considered their oppressors. That struggle had resulted in the upset of the lopsided balance of power. Now the fluid quantity of public opinion was being called in to redress the balance, in some different way.

On the face of it, the commission before which Mitchell appeared was far from being packed for him. The chairman was Judge George Gray of Wilmington, a distinguished jurist, a representative of the country at the Hague Peace Conference—which had not brought peace. The recorder, almost immediately made a full member, was Carroll D. Wright, the United States Commissioner of Labor. The eminent engineer required was General John M. Wilson. Bishop John L. Spalding, added by President Roosevelt, was an eminent Catholic prelate and scholar. Edgar E. Clark, the "eminent sociologist," had been a union official. Edward W. Parker, who was later to be editor of the operators' journal, and Thomas H. Watkins of Scranton, who had been an independent operator in his younger days, were generally held to be hostile to the union. Bishop Spalding, Wright and Clark were held to be sympathetic, although some labor leaders objected to Wright. Judge Gray and General Wilson were somewhat unknown quantities.

Mitchell, whose appearance created a stir in the public hearing, stepped right into the major argument. Without the union the men would be bulldozed; any wage increases they might gain would be taken away from them under cover, bit by bit, for they would have no spokesman to defend them. The essential thing, therefore, was recognition of the United Mine Workers. This had been part of the controversy with the operators ever since they saw that the strike was serious. At the first Washington meeting they had made it plain to Roosevelt that Mitchell was out of their picture. Now they immediately picked up the challenge which Mitchell made, attacking the union as an alien organization, a means by which the bituminous operators could disrupt the competing anthracite industry. The frequent occurrence of local strikes in the bituminous fields was held up to indicate the inability of the union to discipline its own members, its lack of discipline making it an untrustworthy and undesirable part to any contract.

Mitchell did not get off the stand until he had undergone four days of wracking cross-examination. In the course of this he not only gave the logical case for the union to the country; his personality as well as his clear testimony was the embodiment of the significance of that case. There was a sneer in the question addressed to him:

> "So that you are absolutely incompetent to inform this Commission why we should be required to pay our employees $600 a year except your sentimental opinion that a man ought to have $600 a year. . . ."

They talked of liberty to a man who had known for the greater portion of his life just how dependent people are on the employers. They asked:

> "Do you mean to say that a father working in that line who believes that he has a son that may be President of the U. S.—and your father had a son that may be—is not at liberty to work as long as he chooses in order to help that son to the position you occupy today?"

He retorted:

"Mr. MacVeagh, your company regulates that now by making a day ten hours . . . There is no disagreement about having a maximum of hours . . . We have disagreed as to what the maximum should be."

He showed an unexpected solemn wit. MacVeagh was cross-questioning him on the crime of calling strike breakers "scabs." "Having designated one as a scab, what else would you do?" Mitchell smiled: "That is all." MacVeagh spoke of a miner who owned forty thousand dollars' worth of property. Mitchell answered that he "must have been one of those who mined the miners." MacVeagh showed concern about putting the cost of increased wages on the bowed backs of the poor through a raise in the price of coal. Mitchell suggested an alternative to the operators: "They might take it out of their profits and so put it on the bowed backs of the rich."

After this, the examining counsel, who was described as a "legal giant," remarked to him: "Mr. Mitchell, you are the best witness for yourself I have ever faced in my life."

During the week of adjournment to prepare figures, the counsel, Wayne MacVeagh, who had been attorney-general of Pennsylvania, made an effort to have the operators settle with the miners out of court. His attempt failed, according to Lloyd, because George F. Baer wanted a "vindication" as well as the opportunity to present evidence on violence and boycotting.

Then followed days and days in which the witnesses on the miners' side, workers, wives, children, mothers of miners testified. Maimed men who had never received a penny of compensation for their injuries—one of whom testified that a miner was not considered injured unless he lost an eye or leg—let themselves be seen. They were followed by widows of men who had been killed and who were now being charged year after year with company house rentals. The case of Andrew Chippie, twelve years old, whose meager pay of forty cents a day was credited against the debt left by his father, killed four years before in a mine accident, was as strong an example of industrial servitude as that of James Gallagher, who worked at one stretch for seventeen years and six months without ever having drawn a cent of cash pay. The case of Henry Coll, whose only offense in the eyes of the mine owners was that he was a member of a relief committee for the strikers of the philanthropic Mr. Markle, brought tears to the eyes of the audience and the commission. He had been evicted with his sick wife, his one-hundred-year-old mother and his four children—one of them an adopted child whose father had been killed in a mine accident. He himself had not a whole bone anywhere in his body except his neck. The eviction resulted in the death of his wife the night before he gave testimony, and at the time he spoke he could not tell whether his mother was living.

Men, women and children, of every nationality, skilled miners, laborers, breaker boys, engineers on whom depended the lives of the men underground (and who worked as long as sixty hours at a stretch without sleep), witty Irishmen, broken-spirited men—all appeared in procession. One child of eight, who looked five, earned sixty-two cents a week, and because he was too young to know what would happen to him if he did not tell the truth, was not allowed to testify. But his childish figure made its impression.

So terrible was the testimony of the two hundred and forty witnesses that the commission begged that the moving spectacle of horrors be stopped. It is doubtful whether ever before or since, so striking a living picture has been presented of the death toll of industry and of the misery of these men and children who toiled and lived in darkness and cold.

Mitchell had, at the suggestion of Henry D. Lloyd, engaged as chief of the miners' lawyers, Clarence Darrow of Chicago, a man whose reputation as a defender of those who challenged the existing social order was already in the making, who eight years before had defended Debs in the Pullman strike. Four years later he was to run up against William Borah in the defense of W. D. Haywood, Moyer and Pettibone in Idaho. Darrow had some idea of attacking the whole foundation on which rested the prerogative of the anthracite owners. He went to Boston to see Louis D. Brandeis, later appointed by President Wilson to the Supreme Court, then one of the most noted proponents of competition in the country. According to Lloyd, lack of available evidence made an attack on the trustification aspect impossible at the moment. Later, according to one account, when Darrow, after watching the whole long procession of misery, wanted to show that it demanded a new social order, Mitchell stayed his arm. For him the immediate thing was the essential factor. He wanted a case made for the miners, first, last and all the time.

George Baer, president of the Philadelphia and Reading, still shining in the glory of the apostolic succession, had his chance to draw a picture in contrast to that which the miners had painted before the public. The miners had shown what happened to men's lives through the silent, undramatic violence involved when company officials had complete right to say whether a man worked or not, how long he worked, and how much he earned. Mr. Baer painted a picture of violence inflicted on poor, honest men who had tried to work while the strike was going on, of passions inflamed by union agitators. Now people were beginning to tell the difference between cause and effect. The miseries of the miners' lives almost excused and explained away some sharp gesture of retort. Mr. Baer then tried the legal aspects of an industrial situation for which no law had been drafted.

He spoke of the surrender of the rights of an individual in joining a union; accused unions of restriction of output; spoke of the men's violence; took occasion to question Lloyd's knowledge of the industry; referred solemnly, even as employers do now, to the inevitable workings of the eternal law of supply and demand, pointing to the surplus of men in the field as proof of the attractive wages. According to Baer, the only real monopoly in the coal fields was the miners' union. Altogether he made the speech expected of the writer of the "divine right" letter.

It was a bitter stinging attack on those who had challenged the employers' rights to run what they considered their world. It was a valedictory to the hired help. Let them know their places; he knew his.

When the speech was over, while people gasped, Mitchell stepped across and shook his hand. Was it a gesture for the press and public? Had he learned already that people were simple and remembered a gesture when a million words had been forgotten? Was it an insult, a method of applauding Baer as one might applaud the outrageous villain in the melodrama of the day, treating Baer as only a paid official, subject to the approval of the real owners of the industry, now simply earning his salary? Was it something quite else, a tired man's substitution of the wish for the fact, the almost Christ-like desire that the wounds of battle could be healed by a touch and then forgotten? Or did he see himself so close at heart with Darrow and the men of the left, who felt that if ever revolution were justified, it was justified now, that he had to shake himself, definitely remind himself that too many men trusted him for their immediate bread to allow him to become a revolutionist, that he not only had to let the men of the left but himself know by standing up and walking across the room and shaking hands that he could not allow himself the luxury of being one of them?

Years before, in the days in Spring Valley when miners were offering themselves into slavery, Mitchell had heard another president of the union say to his people, "You have reached the limit of human endurance." But Mitchell had lived on with the otners, going without the meat, dividing the bread. There was no limit to human endurance for him and for the men he knew. For Darrow, for the middle-class public who listened to the human wreckage of the industry, for humanitarians touched by some great dramatic and messianic dream, the limit of such endurance might seem to be rising before their eyes then and there. But to Mitchell the miracle was that he who had been nothing might give to these, his people, even a few of the things that would make life more desirable. The simple things that would help them were the things he wanted to do. Anything that would help them was what he wanted to do. He had appeared before the Commission saying that the interests of capital and labor were inseparably allied. Now he shook Baer's hand. The statement and the gesture were both of a piece.

Darrow stood up. He had seen the handshake. What he thought of it he told no one. . . . A month previously, Lloyd had written his wife:

> "Darrow . . . does not like any more than I his (Mitchell's) going to New York to attend the Civic Federation. The little fact that for this visit to New York he bought a derby hat, discarding the black soft felt hat by which he is universally known, illustrates the tendency toward conformity resulting from such association and likely to increase and in the wrong direction."

In later years, a radical newspaper traced Mitchell's "decline" by the evolution from miner's cap to silk hat. Lloyd, secure in the social as well as the economic world, and Darrow, equally at ease and more reckless then as now of his appearance, were puzzled by the purchase of the hat. They would have been puzzled by his inner lack of ease at the formal dinners of the Civic Federation, at the White House, at press banquets. In all probabilities it was Miss Morris, anxious that every detail be correct, who had suggested the purchase of the hat. . . .

Upon Mitchell's insistence, Darrow had already deleted the most inflammatory parts of his summary to the commission.

What there remained in it of dynamite he did not squelch. Baer he swept aside:

> "If John Smith earned $300 a year, it is no answer to say that Tom Jones murdered somebody in cold blood. . . . It is no answer to say that someone's house has been burned. . . . The question is what has he earned."

He showed that Baer's figures, analyzed for the commission, told that forty-nine per cent of the workers got less than $200 a year; that nine per cent earned between that and $300 a year; and that only five per cent got over $800. He showed that the laborers, taking the highest possible average, did not get more than $333. (The final report of the commission gave the average for all workers, including the "composite" miners, as $377 for 1901—a year of unusually high earnings.) Of these laborers, for whom Mitchell had insisted that the union be on an inclusive industrial rather than a craft basis, Darrow commented:

> "The lawyers are in front of the miners, and the miners are in front of the mine workers. Most of the people we have had on the stand are contract miners. The poor devil who loads the coal, we do not want to forget him. What has he been getting? He works in the ground. The falling rock, or a stray car or a belated blast will catch him the same as the experienced miner. His business is almost as dangerous as the contract miner. More than five out of every thousand of his craft are killed every year, to say nothing of the maimed and the crippled and the blind, who are turned out under the beneficent law of the state of Pennsylvania . . . to the almshouses and highways and byways . . . because . . . I believe there is not another state in the union where it is as difficult to recover compensation as in this commonwealth of Pennsylvania.
>
> "When I think of the cripples, of the orphans, of the widows,

of the maimed who are dragging out their lives on account of this business, who, if they were mules or horses would be cared for, it seems to me that this is the greatest possible indictment of this business that can possibly be made. . . . There is not any record of how many accidents there are. . . . An accident must be pretty serious to be recorded. As our old friend Gallagher said, you do not count it an accident in the mine unless you get half killed. . . .

"If forsooth, these poor miners are to have shorter hours or more pay, nothing short of a calamity will overtake the industry of the east. . . .

"If the civilization of this country rests upon the necessity of leaving these starvation wages to these miners and laborers . . . if . . . it rests upon the labor of these poor little boys, who from twelve to fourteen years of age, are picking their way through the dirt, clouds and dust of the anthracite coal . . . if it is not based on a more substantial foundation than that . . . it is time that these captains of industry resigned their commission and turned it over to some theorists to see if they cannot bring ruin and havoc a good deal quicker. . . ."

During the four months of hearings which followed the five months of strike, the miners held a victory convention at which Mitchell was acclaimed by his own people. There was serious talk of running him for the presidency of the A. F. of L. He stopped it. He wanted no political combat. He was in the work he loved. In the joint conferences that year the bituminous miners won another substantial increase. For the anthracite miners it now became a matter of waiting. The fate of the men was now no longer in their own hands, nor in Mitchell's hands, but in the hands of those who represented that quantity little known in the mine fields—outside public opinion. The hearings were over, but the press was not closed. A man could still say in public what he wanted the commission to remember. Mitchell went to Chicago on February

16th. Six thousand people swelled a reception at the Auditorium. Eleven hundred labor men were seated on the platform. As he entered the band played "Hail, the Conquering Hero Comes." Lloyd introduced him, "first in strikes, first in arbitration, first in the heart of workingmen." Eight years before Debs had been welcomed there by Lloyd, after leaving jail for his leadership of the Pullman strike. A few years later Chicago was to become one of the key centers of the Industrial Workers of the World, men who declared that there was no possible truce or compromise between the employers and the employed class.

To this audience Mitchell repeated that there was no irreconcilable conflict between labor and capital. Righteous rage against the coal operators was missing from his speech. As if it were a question of absolutely killing any doubts in the minds of his audience as to where he stood, he added: "We are all poor simply because we cannot get rich honorably." Even that they applauded.

What did he gain by swinging to the right so forcefully? Did he think that what the workers needed most, next to a little more food and freedom than they had, was such self-respect as might come from the sensation of being considered partners with the men who owned and ran the country? Or had he been flicked on the raw by the accusations that he had been flattered away from his original purpose? Or had he set his face a little further ahead, hoping to make the romantic dream complete, hoping to give in his own person an evidence that the miners' place in the country was now changed for all time? He was not unaware of what was being talked around.

Almost a month before this meeting the chairman had written his wife:[1]

> "I got a very important piece of news late last night from Mitchell on the dead q.t. You must never breathe a word of it. An intermediary has lately come from Teddy to Mitchell, telling him he need not worry about the findings of the commission. The President is worried by the talk of putting JM into politics. The Democrats are actually proposing to give him the Vice-Presidential nomination."

The Chicago speech did not make it impossible for him to be considered in that way. But those who knew him accepted his statement that he could think of no greater honor than to be president of the miners' union.

The decision of the commission was announced on March 22, 1903. Was it victory? On the face there were some gains. The union had asked for a twenty per cent increase on piece rates, an eight-hour working day, the establishment of a standard of twenty-two hundred and forty pounds to the ton, no discrimination for union membership, and an agreement with the United Mine Workers providing for a method of adjustment of disputes. The award provided for a ten per cent increase in wages and a ten per cent reduction in hours (a nine-hour day). It declined to fix the method of payment of the standard for the number of pounds to the ton, but gave the contract miners the right to elect and pay checkweighmen and check-docking bosses to look after their interests—if a majority so desired. It contained provisions against discrimination in the distribution of cars and in other ways. It provided

[1] Unpublished letter (courtesy of Caro Lloyd).

for direct payment to helpers. It reëstablished the sliding scale, by which workers were to get increases in pay with increases in the price of coal. In place of the express recognition of and dealing with the union, the award called for the creation of a board of conciliation. This was to consist of six persons, three to be the representatives of the organizations of the majority of the workers in the three districts, and three to be appointed by the operators from the three districts. If these six could not agree, an umpire was to be appointed at their request by one of the circuit judges of the U. S. Judicial Court. It was this board and this umpire which were to take up the many grievances and disputes which arose between worker and operator—after an attempt had been made to adjust them between worker and employer directly.

Mitchell took what comfort he could out of it. He said: ". . . The organizations of the three districts meant, of course, the District Organizations, 1, 7 and 9, of the United Mine Workers of America." The presidents of these districts were appointed as representatives upon the board of conciliation. The operators at first refused to recognize them, but finally, after stoppages and stormy sessions and the confirmation of the men's election by a special convention, the three district presidents were recognized by the board. The award was to continue for three years—to April 1, 1906.

Criticism of the award was plentiful; men like Debs considered it proof of Mitchell's poor judgment. Mother Jones, in later years, insisted that Mitchell had been carried away by the honor and glory of being patted on the head by President

Roosevelt. The miners themselves, to judge by their mass meetings, accepted the award with mingled feelings although, on the whole, satisfaction was predominant.

The miners got some ready cash because the award started from the day they went back to work. That helped. They saw themselves for a little while as a great union, come through a serious test with flying colors, able to boast of the "greatest labor leader" of them all.

Was it victory? Had the clashing nationalities been wrought into a workable league of nations? Had the conflict between the craft workers and the unskilled been integrated? Had leaders developed who could carry on when he went back to the task which the union held for him throughout the scattered bituminous fields? Or had he allowed himself to become too great for them, not only a symbol that bound them, but the only symbol that bound them? These were the questions not only his opponents but he himself asked. In the five years that followed, he was often discouraged by his belief that he had failed here. It was only after his resignation in 1908 that he saw the fulfillment of the promise the anthracite miners had made to themselves and to him in 1900 and 1902.

VIII

THE FABIAN INTERLUDE

MITCHELL did no great amount of basking in the warmth of his own people's approval. At the end of the anthracite strike he told Weyl, "I am almost afraid of this victory. I can never be what the people will now expect of me." He knew his own limitations and the limitations of the situation. For the moment, however, he met the unprecedented expectations placed in him. At the Joint Interstate Conference in February, 1903, a ten per cent increase was obtained for the miners of the central competitive area, followed by a like gain for the outlying unionized districts. This was later said to bring their increases in the seven years since 1897 to sixty-six and two-thirds per cent in some cases. On the average, it was certainly much higher than the amount granted by the anthracite commission in that field. There were other gains. Michigan, Tennessee and Central Pennsylvania succeeded in improving working conditions. Even West Virginia, where the strike had been lost, showed gains in certain sections. The joining of the four southwestern states, Missouri, Kansas, Arkansas and the Indian Territory, into an interstate joint conference such as that in the central competitive area was hailed as an advance. In Alabama, Judge Gray, who had been chairman of the Anthracite Commission, was chosen as arbitrator and prohibited the employment of boys under fourteen. In Illinois, through the efforts of the union, the State legislature set the minimum at sixteen. In Montana all the miners were in the union. It crossed the lines over into British Columbia. Including the anthracite miners, there were about three hundred thousand card-carrying men.

These men were scattered widely throughout the country and any strength they might develop in obtaining better conditions was dependent to a large extent on their success in bringing into the union the remaining one hundred and ninety thousand employed in the industry. It began to be a problem in large scale salesmanship. Against the urgent advice of his friends, Mitchell placed Tom L. Lewis who he knew wanted to succeed him, in charge of the staff of sixty organizers who had previously been under his direct supervision.

Then in 1903 a depression set in throughout the country which affected the coal industry. By the beginning of 1904, Mitchell was faced with the resultant problems. A strike begun in Colorado was in a bad way. It looked as though the moment had come when he would have to prove whether he had ever been more than a fair-weather leader. In Boston, Samuel Gompers was telling the convention of the American Federation of Labor that it was the height of economic unwisdom for anyone to attempt to curtail the consuming power of the masses and recommending that the working people resist any attempt to reduce their wages. It became clear, as 1903 drew to an end, that the coal operators were going to ask the union to take a backward step. They pointed to reductions which a hundred thousand steel workers had taken, reductions in glass works and type foundries. From the miners letters poured in that they would rather strike all winter than accept a cut.

In this atmosphere the Joint Conference of the Miners and Operators of the four central competitive states opened in the last week of January, 1904. The miners this time asked for no increase in wages. They did ask for the extension of the payment on the run-of-mine basis to Ohio and Western Pennsylvania, and further, that men mining coal by machine be paid an additional seven cents per ton, so that they might earn as much as the pick miners.

F. J. Robbins, of the Pittsburgh Coal Company, whose company owned mines in all of the four states, and who was a member of the National Civic Federation which had been informed by Gompers that labor would resist all reductions, acted as spokesman for the operators. The time, he said, had come to call a halt. No arrangement was a good arrangement which did not work both ways. There was disorganizing competition from the non-union fields, much bankruptcy among the operators in the central fields, depression throughout the industry. The operators wanted a reduction of fifteen per cent to last for two years. He announced that the operators meant business when they presented this demand.

What was the responsible leader of the union to do? His name had everywhere been identified with success. What answer could be made, he did make. It seemed to him a strange philosophy that a reduction of wages would relieve an industrial depression. The operators themselves would not be benefited by such reduction. The railroads would have immediately taken away from them all advantages. The cost of living had not gone down. Why make the miners bear the brunt of hard times? Was there really a depression? Where the claim had been advanced in the steel industry a few months back, the furnaces were now operating at full blast.

To this he drew the retort from Robbins that the operators would accept seventy per cent of their profits for the past year if the miners would guarantee them, and the miners should take everything over that. It was not accepted as a great philanthropic bit of profit-sharing. Mitchell was curt: "We will not, of course, accept any propositions of that kind." Then he pointed out that Mr. Robbins' company had not only paid a seven per cent dividend, but had accumulated a surplus that was good for that same percentage for the next five years. There was a wrangle about that. Meeting after meeting of sub-committees, held in secret session, brought the operators' demands for reduction down to twelve per cent. Then there was deadlock, danger of a definite breaking off of relations.

Largely through the good offices of Mitchell's friend, H. N. Taylor, the final break was averted and the conference adjourned until further call, with a special committee elected to consider the matter. The miners held a special convention. They were in no mood to take the reduction. The national treasury contained a million dollars, the Illinois treasury three-quarters of a million, the Iowa district had in two years accumulated a hundred thousand, Ohio had still more.

Was there an irrepressible conflict, or was a working arrangement whereby both employers and miners could be sure of some existence the essential that could not be foregone? It was again a presidential year. Much had been made of that circumstance in 1900. Would the Democrats renew the rumor begun in 1902, this time with the double threat of a Bryan to carry the agricultural West and a Mitchell as team-mate to carry the industrial East? And would the man who was trying to formulate a policy for the miners under these trying circumstances seriously consider such a possibility? When proffered nominations in the past, he had always replied that he could think of no higher position than the one he held.

It was not yet the day when people spoke or wrote at length about the absence of a working class in the country. Mark Hanna, a month earlier, after saying that organized labor had come to stay and praising the miners' union in particular, had written frankly what he considered the situation: "What is the public? In our country it is made up of capitalists and laborers. With the exception of a very few, every man in the United States is an employer or employee." If there were only two classes, was it true that there was a constant inevitable conflict between them? There were men amongst the miners who came to that conclusion. But early in February, when Marcus Alonzo Hanna died, Mitchell requested the miners to refrain from work on that day. It was a request which drew forth considerable comment. The Wilkesbarre *Labor Unionist* said: "At this writing one doesn't know how the miners will take the suggestion as many of them are now laughing at the irony contained in the proposition." Just what Mitchell intended, whether it was a gesture of reward for Hanna's considerable services during the anthracite strike, or whether Mitchell had intended it to perform the same service in the public estimation as his handshake with Baer, he never told.

At the resumption of the sessions, one of the operators took occasion to ask for "peace in the name of him who has gone who would rather be the preserver of industrial peace than be president."

The resumption of the hearings brought a more intense repetition of every argument that had been gone over before. There was little new. At one point Robbins asked for arbitration. Mitchell was willing to arbitrate only the price of machine mining. During the anthracite strike, in a moment of depression, he had said: "Now there remains nothing but defeat, victory or arbitration." He knew that arbitration was neither the first nor the second. It was a bad gamble, not to be taken when the system of collective bargaining was made. He showed that in 1903 the union miners in Ohio had earned on an average only four hundred and thirty-six dollars. Mr. Robbins remarked that that was because West Virginia was paying wages which made it impossible for the Ohio operators to compete. Mitchell replied that Robbins alone in the Pittsburgh district was producing twenty million tons to the total of West Virginia's twenty-six millions. On the sixth day of the resumed conference the Joint Scale Committee brought in its report for a reduction of five and a half per cent. The operators all voted for it. The miners' delegates, although their representatives and officials had been on the committee, voted solidly against it.

The delegates had voted. Some new force had to be called into existence to break the hard-drawn line. In the anthracite strike there had been public opinion acting through Roosevelt. Here there was no strike and no aroused public opinion. The miners went into session alone and decided to submit the question to a referendum of the membership. For a second time, Mitchell was going before the miners with a policy which seemed to be against its wishes. In previous years, he had had to confront a body of representative delegates, who also elected him to his office. This time the vote went directly to the membership. (This year too, at his own suggestion, his election and that of the national officers was by direct referendum.) He felt justified in explaining his attitude in a letter attached to the call for the referendum. In that letter he was joined by Tom Lewis, the vice-president, and W. B. Wilson, the national secretary.

The letter is most interesting in the way, on the surface of the matter, it denied practically all of the statements Mitchell and the other officers had made in the public hearings of the joint conference. It was affirmed that the union had never been afraid to strike where there was a possibility of beneficial results. But then it agreed that since 1897 the organization had ridden on the crest of prosperity. Where Mitchell had scoffed at the idea of depression the letter remarked: "It is conceded . . . we are slowly moving toward an industrial depression." Where they had talked of the importance of keeping up the consuming power of the workers to end a depression, they now said: "The events of the past have shown conclusively the fallacy of striking when the markets are going down." Where they had said West Virginia was unimportant because of its relatively small production, they now remarked: "No chain is stronger than the strength of its weakest link. We have some weak links in the chain . . . West Virginia outside the Kanawha fields . . . the Irwin, Connellsville and Klondike fields in Western Pennsylvania cannot be depended upon to cease work if a general suspension is declared. . . ." The fact that it was a presidential year they gave as a reason against the strike as such years "are proverbially dull because of commercial disturbances." The objection of their own men that the decrease was to last for two years instead of one was met by the statement that "two years will give you time to organize in unorganized fields." It was also pointed out to those who thought that in time public sympathy could be aroused that "you would not have the sympathy of the public in a difficult and prolonged strike."

There was truth in this letter and there had been truth in what the officials had said to the operators. They could go before the membership sure that it would be understood that they had put up an honest and hard fight. This letter, which bore only one side of the argument, went to each local. It ran into a strong opposition which, however, had no official funds to circulate the locals in its interest even if it should consider that advisable. The official letter, which mentioned no action by the National Executive Board, was seconded at length in the columns of the *United Mine Workers' Journal.* It could say what the officers could not well do themselves: "The man has never drawn breath that can truthfully point to a single tactical blunder of John Mitchell, T. L. Lewis, or W. B. Wilson."

In the course of the argument, Mitchell's picture was turned to the wall in his own home town and elsewhere. There were men in the field who had known him for a long time and who never spoke to him again after that, even when he entered their own homes. His family as well as he felt the brunt of his action. The fighting was carried from the local halls into the saloons.

When the vote was finally cast, there were sixty-eight thousand for the strike against any reduction, and one hundred two thousand against the strike for the reduction. A switch of eighteen thousand votes would have brought a strike. In Illinois not only the Socialists but Mitchell's old friend Ryan, who had led the opposition in the convention, rolled up a large vote for the strike. In Mitchell's home town of Spring Valley, only one man in sixteen approved of his position. Feeling in Illinois was high against Mitchell for other reasons. Under a peculiar arrangement in that state as in the southwest, he acted as impartial chairman in disputes between the district organization and officials. This was a rôle admirably suited to his temperament; he was impartial and miners were puzzled and angry to find their national president occasionally deciding against their district officials. A man with whom Mitchell had worked in the mines addressed him as "Mr. Mitchell." Mitchell, hurt, insisted on an explanation and learned that the man was doubtful about a leader who banqueted with captains of industry.

While the miners accepted the reduction as a bitter pill, the public press lauded Mitchell to the skies, proclaiming it the greatest act of his career, proving his sportsmanship, the union's willingness to give and take, the wisdom that did not balk at temporary defeat. But at the same time, other explanations were evolved to account for Mitchell's attitude.

One was very simple: there had been collusion between Mitchell and the operators. Another was that the operators and Mitchell's opponents in the union were combining to get him out. According to this story the operators hoped for his defeat in the referendum and the subsequent break-up of the central competitive field; his opponents in the union hoped to take advantage either of his recommendation or of his defeat in the strike to force him out, or to lead him to accept the appointment as United States Commissioner of Labor which, it was rumored, was to be offered to him. The fact that Lewis' name had been signed to the letter, although the signature of the vice-president was not requisite, seems to destroy the force of this explanation or else to show that Mitchell forced Lewis into an open stand.

The immediate results of the reduction were bad. In May, 1904, Lewis reported from the field that along with the reduction there was a great deal of unemployment. Strikes were being carried on in Western Pennsylvania where certain operators attempted to go beyond the agreed reduction. Eight thousand men were out in Ohio for the same reason. A reduction on the same basis as the competitive field had to be taken in the southwestern field, after a referendum in which thirty-four hundred voted for a strike and fifty-nine hundred against it. In Tennessee a seven per cent reduction had to be accepted.

A new feeling of doubt and hesitancy about his leadership met Mitchell when he returned to the less dramatic and now more difficult work of extending a union that had been widely advertised as an organization willing to take a reduction without anything more, apparently, than a fight of words.

One of the worst situations was developing in Colorado. In the spring of 1903, Tom L. Lewis and the district president, William Howells, had begun an organization campaign. There were about eleven thousand men in the field and about fifteen per cent were organized. Most of the union membership was in the small northern area. The more important southern field where the Rockefeller company, the Colorado Fuel and Iron Company, and the Victor Fuel Company, Gould interests, controlled the production, was almost completely unorganized. Conditions in the south were comparable to those of feudal West Virginia and of the anthracite miners prior to the 1900 strike. The operators controlled not only the working conditions of the men, but owned their houses, the stores where they were compelled to buy, the physicians for whom they were docked, and the local police forces and courts.

In August that year a manifesto to the Governor and the people of the state recited the grievances of the miners and the refusal of the operators to meet with them. At a conference only three companies from the northern field responded. The southern coal companies paid no attention. In September a district convention was called to formulate demands. Mitchell was not there; neither was Lewis; but Charles Moyer, President of the Western Federation, addressed them and spoke in favor of a strike. His organization which, over the protest of the United Mine Workers, included some coal miners, was known to be in opposition to Mitchell and his whole attitude of conservative advance, and its denunciation, particularly in the utterances of William D. Haywood, its secretary, was couched in no uncertain terms. At this time it was conducting a strike of the metalliferous miners of the state and wanted the coal miners out in support.

After the demands, similar to those in the anthracite field, had been formulated and a vote favoring a strike taken, the district officials traveled to Indianapolis to secure approval of the strike by the National Executive Board. They insisted "the time is now or never, and if you don't strike now in Colorado, you will set back the work of the organization ten years." The National Board deliberated, invited the two large southern companies to a conference, receiving a refusal from the Colorado Fuel and Iron on the ground that their men did not want the strike and that the union did not represent them. The Victor Fuel Company refused a conference but wrote of raising rates from ten to twenty per cent. There was hesitation. The industrial depression, of which the operators had made so much a few months later, had already set in. Colorado had always been a difficult state; it was hardly in the first ranks of importance as to production and competition with the main field, although the treatment accorded the workers there had been a crying outrage for some time past. Perhaps the Board might have counseled the miners to wait for a better time, but their hand was forced. The Colorado State Commissioner of Labor arrived in town and requested permission to address the board. He asked for an opportunity to avert the strike and to secure an eight-hour day. He left the office and, without any authority, wired to Denver that there would be no strike. The National Board could not be put in the situation of endorsing an agreement it had not made and of being rushed over the heads of the miners in the district into opposition to the strike. They endorsed the strike, and it began on November 9, 1903.

The operators in the northern fields immediately hurried into conference with the union officials and offered an eight-hour day and a slight increase in pay, contingent upon the miners' obtaining similar conditions in the south. The miners saw no opportunity of doing that without a strike, and, on a referendum, rejected the offer. Although the card-carrying members were few, the unorganized workers poured out in response to the call. The northern operators made another offer, again rejected. The miners did not want to desert the south. The district officials and organizers in charge, including Mary F. Jones, already famous as "Mother Jones," were against this offer. She was vehemently against it. Already in her sixties, she had seen service in the West Virginia and anthracite strikes, and was to come upon the scene of many great struggles in the next twenty years, making perhaps her last appearance years later at the side of President Coolidge, whom she endorsed. Then a fiery agitator, called a "stormy petrel of labor," mob leader by temperament, with a physical courage and a vocabulary that awed men, she was an important force at mass meetings, and men went far to listen to her. She had expressed open disfavor with Mitchell's policies at times, and while he did not agree with her methods, he recognized her influence and ability as an agitator and kept her in the field. He had a capacity for appreciating people who were unlike him, enjoying the hearty Falstaffian mirth of men like Loftus and Taylor, admiring daredevils like Mother Jones and Alex Howat, later to be expelled by another president for his wildcat strikes. Howat and Mitchell had grown up in Braidwood together, and the shy, repressed boy never forgot his envy of the wild youngster who would leap from a coal pile no matter how high it was or how perilous the landing. He bore no malice toward the employing class, but he could laugh with the simple laughter of that boy when Mother Jones or Howat told stories of having embarrassed or infuriated coal operators. He could not bring himself to denounce or expel them, even when his board demanded it, lest he give comfort to that employing class.

A little later, on November 21st, the operators asked the northern miners again to settle. The strike had been going only thirteen days. The representative of the National Executive Board produced wires from the national office, urging acceptance of the offer, which was not now contingent upon similar gains in the south. To Mother Jones and others it seemed an outright betrayal of the southern miners. A vote taken at a mass meeting showed two hundred twenty-eight for continuing the strike, one hundred sixty-five for returning to work. This was less than a third of the men on strike in the north alone. The national office asked for a secret ballot, which, when taken, showed a complete reversal: Four hundred eighty-three voted to return to work, and one hundred thirty against it. By the end of November the northern men were back at work, every demand granted except the check-off.

A struggle in the southern fields continued, more bitterly than ever. One of the northern local leaders, Robert Randall, claimed at the 1905 convention of the miners that had the northern miners stayed out, "it would have meant a victory for both the coal and metalliferous miners of the West. . . . The day the northern miners returned to work, the discouraged southern miners were defeated, and the strike was lost. The Denver Citizens' Alliance, the Victor Fuel Company, and the Colorado Fuel and Iron Company, with the able assistance of John Mitchell, won a victory over the miners of the West that has forced them into a state of abject peonage and closed upon them the door of hope for years to come."

Almost immediately the two large companies began evicting strikers from their houses. The strikers set up their tent colonies. Communication between them was made practically impossible. Union men were jailed on entering a town and escorted out by force. Wholesale arrests, forced marches over the mountains, deportations, riots and killings became everyday affairs in Southern Colorado.

The situation became so serious that on December 2, 1903, Mitchell himself went to Denver to consult with Governor Peabody. While there, he addressed the Chamber of Commerce at their invitation, outlining to them the purposes and policies of his organization. Randall described the circumstances of the meeting: "Nineteen thousand men, women and children, cold, hungry and dressed in gunny sacks (were allowed) to wait encamped on the barren hills and in the desolated canyons of Colorado, New Mexico and Utah . . . while you discussed the labor question over the banquet board with Civic Federationists and (the) Denver Citizens' Alliance." Before that meeting, Mitchell had stopped off in Trinidad, the center of the southern strike, and had, according to Randall, said to the strikers: "Be peaceable, be law-abiding, and strike, strike, strike until you win if it takes five years."

Mitchell's version of the Trinidad meeting was somewhat different. "Delegate Randall says they all knew their cause was lost when the northern men went to work. . . . How do you suppose the men in Trinidad met me? . . . The train on which I went . . . was five hours late. For these full five hours, thousands of men, women and children waited at the depot for the train to come in. Were you there? . . . Did they receive me with clubs and guns? I have received about as enthusiastic receptions as most men, but I never received a more enthusiastic one than the people of Trinidad gave me. . . .

"I said to them that while personally I favored other means than strikes for the adjustment of grievances . . . that when all these peaceful means failed, that I favored striking, and striking and striking until they won. I never said they ought to strike for five years. I am not in the habit of making exaggerated statements. If I advised them to strike for five years, that meant we would pay them to strike for five years."

Mitchell had no success in the conference with Governor Peabody, who instead insisted that the men who wanted to mine coal must be protected. That very week two strikers were killed by the militia. Two National Executive Board members, sent by the national office, were assaulted on the train en route to Trinidad and the recovery of one was considered doubtful. That was only the beginning. In March troops were sent in and a state of insurrection and rebellion declared. Another national organizer was severely beaten up, and another striker killed. Every house was searched for firearms, and very few were found, but helpless foreigners were thrown into jail. Press censorship was established. The accusation was made that two hundred thousand dollars had been subscribed by the various coal companies for the carrying through of this military régime. On March 26, 1904, four organizers, including Mother Jones, were forcibly deported, and another eight the following week for no reason other than that their "absence was better for the people." The beginning of the feeling that brought the I.W.W. into the field a year later, that led to the Ludlow killings in 1914, and the shootings in 1927, was in evidence.

Meanwhile Mitchell had gone East again, had been in conference with the operators, and had recommended the reduction. A little later he asked President Roosevelt if there was nothing he could do to help. The President did send a special investigating committee into the field, but it found no leverage for Federal intervention. On June 2, 1904, after the strike had been going on for seven months, a member of the National Executive Board said that no further funds would be forthcoming from the national office, that "close inquiry satisfied us that the mines were being operated with reasonable success and there was no possibility of winning the strike." It suggested that the district call it off. The district tried to carry on the strike for a few months more, but the heart and sinews were out of it.

At the 1905 convention of the union, Robert Randall succeeded in making a scandal out of the conduct of the Colorado situation. Mitchell had been to Europe and had come back in the meantime, leaving directly after the National Executive Board had come to its decision. This was used in much the same way as the mine leaders used the absence of Rockefeller from the field where his subordinates had carried out what seemed criminal and illegal actions.

This was the first time Mitchell had had to stand up and fight across the wreckage of an obvious disaster. His own small failure in West Virginia had never drawn attack; it was easily understood and lost in the record of his accomplishments; it was more a personal matter with himself about which he had felt badly. The reduction in 1904 had not led to an open attack on him personally and now in the 1905 convention it was hardly discussed, perhaps because in some sections the men had earned more money in 1904 than in the previous year.

The main charges made by Randall were: first, that a policy of sectional settlements in Colorado was a betrayal of the strike; and second, that after the sectional settlement was made, Mitchell further betrayed the southern miners by withdrawing the support of the national office. The incidental charges were embroideries on these themes: "Mitchell had dined and wined with capitalists who were notorious labor-baiters. He paid more attention to the desires of the Denver Citizens' Alliance, a creature of the Colorado coal operators, than to the needs of the strikers. He persecuted and drove out Mother Jones because she would not submit to his policy. Five days after the national office had withdrawn its support from the Colorado strikers, he permitted himself to be the guest of

Frank Robbins (who had been so instrumental in the reduction) and of other employers at a sumptuous banquet at which he was presented with diamond mementoes of esteem from the employers. While the strikers starved, he went to Europe, presumably as an elected delegate to the International Mining Congress, but actually for his own pleasure and financial profit. The Congress lasted five days and Mitchell stayed abroad two months, writing, with the assistance of Dr. Weyl, articles which he sold for two hundred dollars each, while he was receiving a salary from the mine workers for the protection of their interests."

Randall, who admitted openly on attack that he had devoted his time to collecting funds for the Western Federation of Miners, and that he had gone back to work in the northern fields following the disputed settlement, brought to the floor of the convention some of the black and white emotional quality of his organization. After telling the convention that "Mother Jones' white-haired head will soon be laid at rest; her voice so eloquent to plead the cause of the oppressed will soon be hushed; her heart that beat so warmly in sympathy for suffering humanity will be stilled in death. But when she is laid forever in the grave, no one can ever say that she ever betrayed, that she ever played false, the toiling and disinherited masses who are fighting the battle for labor's emancipation." He drew a contrast between Mother Jones and Mitchell, whom he called "the little tin labor god of the capitalist class," who dined and wined with "the man (Carnegie) who has the blood of the Homestead men on his hands and who had done more to crush American manhood and degrade

American womanhood than any other man in the United States," that "the time is coming when John Mitchell and leaders of his type who have fastened themselves like carbuncles (*sic*) on our organization will be scraped off."

There was argument back and forth as to the responsibility and actual fact. Vice-President Lewis said that it was certainly not Mitchell who had insisted on or initiated the strike, not he who was alone responsible for its conduct. Other men showed that a greater amount was spent by the national office for the eight thousand Colorado strikers than for the one hundred fifty thousand men in the anthracite strike. They charged the Western Federation of Miners with efforts to disrupt the union. They denied, and the payroll substantiated them, that Mother Jones had been forced out. Mitchell had, in fact, kept Mother Jones on the staff in the face of serious charges which she never denied, and over the protest of the Executive Board.

Mitchell stood up solemnly to make his own defense. He thought the settlement of the northern part of the state a good one. He had not complete coöperation from the district. After he had removed John Gehr, National Board member, for dishonesty in communicating information to the operators, the district president had put him in charge of the strike. He denied that he had spent twelve dollars a day on hotel bills in Europe. He denied that he had stood in the way of the southern strikers and then made the vote on Randall's accusations of him an official vote of confidence. He said: "As far as my present official position is concerned, I would be happy, indeed, gentlemen, if it were your wish to turn it over to-morrow to someone else; happy, indeed, to go back to my home and loved ones . . . to have the pleasure denied me for seven long years. . . . Do you suppose the salary I get here, good as it is . . . compensates me for the loss of companionship? . . . Do you think I crave this position? I wish to assure you —and I defy any man living to contradict what I am going to say—that from the first year after my election, I have never asked, directly or indirectly, for any man's support or for a single vote. I will admit that the first year I ran, I made a contest for the office. But if I stay here one year or twenty years, this office must come to me unsolicited and unsought. . . .

"Some of you have forgotten, but I never have, the conditions you had when I first came with you. I do not claim any part of the credit for the change, but I hope the time may never come when my people must go back to where they were some years ago, when I used to see in the daily papers throughout the country . . . accounts of men and committees going around begging bread for the miners; not miners on strike but miners at work. . . .

"Statements made by Delegate Randall and by men like him do me injury. Some men will believe such statements. A lie uttered can never be taken back. . . .

"Now, gentlemen, if one of the charges this man makes is true . . . I say that I am utterly unfit to represent you; I should be deposed and you should put in my place a man whom you believe to be trustworthy. On the other hand if these statements are untrue . . . then I say I have a right to demand of you that you afford me protection. I cannot protect myself against the scandalmongers all over America . . . I cannot reach them all. . . . Unless I get them here or unless I can get into their locals I cannot answer them. . . . Either this man or I should not be a member of this union."

The motion to expel Randall was made by a Socialist, one John Walker of Illinois, and seconded by another, Wardjon of Colorado, the latter stating that originally he had been opposed to the northern settlement. Randall, who said he "had not meant to attack Mr. Mitchell," was given a chance to retract his charges. He refused it and the motion was unanimously passed by the convention.

Underneath the emotional cloud around this attack on Mitchell, the rather simple attempt to claim a leader corrupt because he failed or others failed, the desire of the delegates to protect their leader from scandalmongering, the kernel of a very important issue was exposed. The issue was not, as some saw it, whether men who thought nothing ill of "dining and wining" with the employers were proper persons to lead a movement of the disinherited of the earth; it was, rather, to what extent the union could extend its control of an industry scattered over twenty-six states when its leaders showed this willingness to make fragmentary and sectional settlements. A year later the opportunism of such a policy became the main problem facing the national union.

IX

THE BROKEN HARMONY

IN 1906 three things came together: the agreement in the central competitive field, the agreement in all the outlying districts including the southwestern field, and the anthracite award expired. The way was clear for unified nationwide action on April 1, 1906, and that action, to judge from the stand of the operators, would have to be one of resistance.

Mitchell faced the miners' convention in no high hopes. He knew himself to what extent the prosperity of the country had helped his leadership, to what extent his conservatism had made the operators willing to deal with him and also, perhaps, had made them sure that he would do anything to avoid a fight. He knew to what extent his strength lay in the unified force of the miners behind him, in his ability for some years to keep factional disputes and differences from policy from becoming public property. For the first time since he had been president, the vote cast for him was less than that cast for Lewis as vice-president. He had been hurt at the time by the distrust expressed in the turning of his picture to the wall. He had been wounded and angered by Randall's attack, he who had tried to rule by affection rather than arbitrary power, to make the men feel the loyalty to the union which he himself felt. He was dismayed now at the open discussion of the extent to which the known opposition of Lewis would influence the complicated and none too rosy situation in 1906.

Personal troubles: the death of another child, a six-year-old daughter of whom he was particularly fond, the loss of all his savings of two thousand dollars in the failure of a local Illinois bank—these too had their share in contributing to his anxiety.

He had gone to the anthracite field during the summer preceding the convention, and had rallied the men, discouraged, in spite of increased earnings and a shorter working day, by their treatment from the Board of Conciliation, their helplessness in view of the umpire's decision that the operators' privilege of dismissal could not be touched. They responded with a new enthusiasm. In the Scranton district alone, ten thousand men rejoined the union. Mitchell rode down the street at President Roosevelt's side. There were Mitchell parades, Mitchell songs, buttons, flags and babies. But now he was getting to know the value of such acclaim, the fate both of men who want leaders to worship and the leaders they worship. He said: "John Mitchell can't save you unless you put forth an effort to save yourselves. . . . 1906 will soon come around and what are you going to do? If you are not organized, I won't work for you. You are not going to do with me as you did with the lamented John Siney. . . . I am not satisfied with conditions here. The hours are too long and the wages are twenty-five per cent lower than they are in the bituminous fields." The fate of John Siney, anthracite leader of the seventies and eighties, who died, desolate, rejected in a period of depression by the miners whom he had organized, haunted Mitchell. He himself, it was rumored, was no longer well.

Roosevelt, Bishop Hoban of Scranton and Mitchell

His speeches that fall were marked by hoarseness and signs of fatigue. In a few months, however, he regained some hope from the fact that the anthracite membership had doubled and was again over eighty thousand.

When he called the miners' convention to order on January 16, 1906, he looked tired. Stories of his illness had been going around and he met some cynical smiles about it. The fact that the three agreements terminated at the same time seemed to the outside world a source of strength in case of strike. He knew well enough that so far as negotiating three new agreements was concerned, it meant a great deal of detailed work and talk at the same time, with very few men doing most of the work and carrying all the responsibility.

The convention agreed on ten demands for the bituminous miners: the admission of the southwestern states and "all outlying districts whose operators are willing to participate"; a twelve and one-half per cent general increase; a universal run-of-mine system of payment; a seven cent differential between machine and pick-mined coal; uniform outside day wage scale; abolition of employment of boys under sixteen years "in or around the mines"; a yearly contract; an eight-hour day; payment for at least two hours' work if the men were required to report for work. So far as the anthracite fields were concerned, the decision of the three districts had been to leave the matter of negotiation in the hands of a committee headed by Mitchell. A conference with the anthracite operators had been arranged for February 15th. Nothing further could be done until Mitchell and the others had attended that conference.

Since a fight seemed inevitable, it was decided that the delegates from all the districts whose agreements expired March 31st were to remain in Indianapolis. In this way they could meet with their delegates at the joint conference at the end of each day and decide on further action. The anthracite delegates were free to leave. The southwestern delegates were instructed to proceed with their joint conferences concurrently with the central competitive field.

The battle began. The first demand—for the inclusion of Kansas, Missouri, Arkansas, Texas, Iowa and the Indian Territory, in one joint conference with the central field—was abruptly refused by the operators in the central field. The miners were taunted with their inability to bring the West Virginia operators to the conference. The operators had no desire to have their own ranks split by the admission of small well-unionized sections and operators. They knew that there was much dissension in the union, and had no willingness to give Mitchell even the paper credit of extending the influence of the interstate agreement.

Mitchell, who was well aware not only of the dissension, but of the fact that the period since 1904 had brought failures in Pennsylvania, West Virginia and Alabama, presented the miners' demand for an increase very simply. The policy of reduction, he said, had been of benefit neither to the miners nor to the operators, no, not even to the public. As he had predicted, the railroads and large industrial consumers of soft coal alone had benefited. The prices on domestic coal to the consumer had gone up rather than down. Nor had the reduction brought the increased amount of work promised. In short, from every point of view, the policy of wage reduction had failed in every respect, except in the sense that it had demonstrated its own weakness. In the pig iron industry, which had been named by the operators in 1904 as the barometer of industrial conditions, there had been an increase in wages. There was prosperity in other branches of industry. The miners, it followed, were entitled not only to the return of the 1903 scale, but to an increase over and above it. In connection with the demand for the run-of-mine payment, Mitchell, in addition to the usual arguments for its essential fairness and practicability, turned the operators' guns on them by showing that most of the local disputes which they mentioned as examples of the lack of discipline, came from a failure to initiate such a system of payment.

Robbins, who was again spokesman for the operators, gave an answer equally brief and simple. He quoted from the union *Journal* to show defeats and lower prices in partially organized fields such as Central Pennsylvania. He denied an increase in coal prices. Competition from the non-union districts made ridiculous the demands for run-of-mine and increased differential for machine-mined coal. What he suggested was the scale which would, according to him, permit more work in the union fields. It was better for the men to have lower rates of payment, more work and higher annual earnings. "There is no buncombe in the position of the operators." He offered no guarantee of this increased opportunity to work and drew Mitchell's retort: "As long as I have any recollection I have heard that old story. . . . A few years ago the miners were foolishly taken in by such a promise. We don't want steady work. We know that steady work is impossible."

The conference went on, with a mutual measuring of strength, for some days. The only agreement arrived at was one prohibiting the employment of boys under sixteen underground, and boys under fourteen for surface work. But even this was not accomplished without the unprecedented move of Patrick Dolan, president of the Pittsburgh district, who spoke against this reform. But he stirred another and greater uproar among the miners when, on the vote to continue the 1904 reduction for two years, he voted against the miners and with the operators. It was not only treachery to his own delegation, it was an open advertisement to the main operator in his district, Robbins, who had moved the continuance of the reduction, that the split in the ranks had gone far. The record is dramatic:

> ". . . The miners of Illinois, Indiana and Ohio voted in the negative. Mr. Dolan voted in the affirmative for the miners of Western Pennsylvania but the other delegates from that district voted in the negative.
>
> "The Chairman declared the motion lost.
>
> "Mr. W. D. Ryan: How did the Pennsylvania miners vote?
>
> "Mr. Dolan: The Pennsylvania miners vote 'Aye.' "
>
> The other delegates from Western Pennsylvania again rose and stated that they had voted in the negative.
>
> "Mr. Ryan: I want to know how the Pennsylvania miners are recorded as voting.
>
> "Mr. Dolan: I have the authority to vote for the Pennsylvania miners, and until they take that authority away I shall vote.
>
> "A number of delegates shouted: Take it away. Take it away!
>
> "Mr. Mitchell: This convention yesterday, by a practically unanimous vote, decided to indorse the action of the miners' scale committee. Therefore the miners vote 'No.'

> "Mr. Robbins: Mr. Mitchell has no right to cast his vote in that way unless he is the dictator of the mining interests of this country. We are here representing states. . . ."

The Joint Conference ended in an uproar. It became clear that if the operators were going to insist upon a continuance of the reduction, there would be a nation-wide strike. When the miners went into special convention by themselves on February 1st, they knew, and the country knew, that they were facing the most serious situation the union had met since 1897, when the central competitive field had been established, and that now they had much more to lose.

While the papers were headlining "Miners Declare for War" and, a little later, "John Mitchell's Fall Predicted," Mitchell bent himself to reëstablish the unity of forces necessary no matter what the character of the final decision was. The very first day W. D. Ryan introduced a resolution which was passed, providing that "no contract be signed in any district until we all get a settlement or go down in defeat together."

This done, the miners proceeded to clean out Dolan. The delegates from his own district passed a resolution of condemnation. The national convention would have ousted him at once had Mitchell allowed it. Dolan's fate was handed over to his own district convention in Pittsburgh five days later. Mitchell was requested to be present but was kept away by the anthracite developments. The man he chose to send there was the vice-president, Tom Lewis. No sooner arrived in Pittsburgh than Lewis was quoted in the public press as having "expressed distrust of Mitchell." Lewis did not run the risk of alienating the Dolan men. The result was that the district convention too ended in an uproar. Mitchell was now informed, as was the public at large, of the extent to which his official family would support him. There was also a story current that Ryan, his old friend and adviser, had split with him on the ground that he had not been sufficiently aggressive. This was denied, with considerable spirit, by both men.

While the press was reporting that Baer, the outstanding power in the anthracite field, and Robbins, the leading operator in the bituminous field, were in conference in an apparent attempt to solidify their forces, and the *Indianapolis News* was repeating the story that the 1904 reduction had been planned as a means of unseating Mitchell and suggesting that the operators had since learned more telling tactics to accomplish that end, Mitchell and the anthracite delegates met with the anthracite operators in New York City. This, the first occasion on which the operators of the anthracite region voluntarily met with representatives of the United Mine Workers to adjust mine wage scales, seemed hopeful to Mitchell.

Then the encouraging news came from Pittsburgh that a second district convention, called to act on Dolan's betrayal of his trust, had removed him from office. In that convention Dolan had made Mitchell the issue. He pointed to the failures in Colorado, in West Virginia, in Meyersdale, Southwestern Kansas and Western Pennsylvania, after more than a million dollars had been spent in these fields. It took "more than a Prince Albert and a red carnation to make a leader," Dolan said. Mitchell had been made by favorable circumstances.

Since the depression he had been proven to be a poor leader. Dolan went further, and stated that at the convention of the American Federation of Labor in November, 1905, Mitchell had told him that he would take the 1904 scale rather than risk a strike. To attack a leader in time of war met with no approval. Deposed as district president, Dolan however still remained a delegate to any special convention of the national organization.

Shortly after this, Robbins, who had seemed bent upon forcing Mitchell out of the situation, and on maintaining the 1904 scale, changed his attitude. Perhaps the outstanding influence in this change was the fact that the United States Steel Corporation, with which he had his main contracts, wished to avoid an interruption in its coal supply. Dolan's defeat may have seemed ominous to Robbins. To some extent he was influenced by H. N. Taylor, the Illinois operator who was Mitchell's close personal friend. Taylor was convinced that an interstate agreement was preferable to uncontrolled competition. He knew too that the acknowledged militancy of the Illinois miners and their large treasury would hold them out until the bitter end, and that the unorganized mines of West Virginia would probably replace Illinois as the second largest coal producing state in the country in the event of a long strike. He pointed out to Robbins that even in the event of a defeat of the miners and a repudiation of Mitchell, the spirit of unionism would not be dead.

There was, however, no excuse to reopen the joint conference which had been adjourned indefinitely. As a result of an agreement to do so between Mitchell, Taylor and Robbins,

Roosevelt was asked to mediate. In answer to his letter, the cause of which was not perhaps known to operators and miners, the conference reopened on March 19th. Robbins now moved to restore the 1903 scale, which meant an increase of slightly less than six per cent. The operators in the states other than Pennsylvania voted against his action, and took occasion to question Robbins' motives and to depose him as their spokesman.

At the same time that this conference was going on, there were special sessions of the miners' delegates to decide on each day's developments. While bituminous miners were beginning to hope that if Robbins could stay in control there would be a restoration of the 1903 scale, the anthracite operators announced a complete refusal of all four of the demands of the anthracite miners. The only proposal they would consider was a renewal of the commission's award. Robbins' deposal and the action of the anthracite operators were blows to peace. Now there were less than two weeks before April 1st when the agreements and the award expired.

Dolan was not as yet completely disposed of, and many delegates objected violently to his being seated in the convention. They were reminded by Mitchell and Wilson that he was entitled to his seat by the constitution, and that they had no more right to violate the laws of the union than Dolan himself. Mitchell pointed out that no one "in the convention has reason to entertain more personal resentment against Patrick Dolan than myself . . . yet I hope that I am big enough and manly enough to lay aside personal feelings when the question of the laws of the organization is brought into question." The district was given the task of expelling Dolan as a delegate, but the convention insisted on discussing its right to expel him as a member. Walker of Illinois asked: "Would it be a violation of the constitution for us to hang him?" and drew the rebuke from Mitchell, "The chair desires to say that such language may not read as well as it sounds. . . . You cannot tell in reading whether such a remark is a joke or in earnest."

The danger to Mitchell here was that Dolan, knowing that his future in the union was over, would, so long as he was a delegate, cause embarrassment to Mitchell by insisting among other things on the original demand for a twelve and a half per cent increase and a strict adherence to the Ryan resolution which provided for national settlements only. Mitchell's action in not expelling him was interpreted to mean that he had Dolan in a position where any action that might disrupt the convention would lead to the end of tolerance.

At this juncture, when stories were being circulated that Dolan had in earlier times turned his union connections to financial profit, Mitchell was suddenly and to his surprise accused, in a very minor way, of the same thing. In the joint conference, the miners were bringing out the connection between the railroads and the coal companies in the bituminous fields, and one of the operators' delegates was charged with being purchasing agent for the Illinois Central. He retorted that Mitchell had "seen the time when you were glad I was an officer in the railroad company." On Mitchell's refusal to let the remark pass, the man explained that Mitchell's family had been given a pass over the railroad. It was not a very serious charge but Mitchell, who was nonplused, would not let it go. He denied asking for the privilege and Ryan arose to explain that he had done it without Mitchell's knowledge. The chairman of the conference ruled that no reflection had been cast on Mitchell's "integrity" but Mitchell was not content. In a statement, rather long for him, he said that the chairman's ruling cleared him only personally, but not his leadership. He wanted complete clearing of the record and of his reputation "before the people of this country," as this was the first time in all the years since he had risen from coal miner to president of a union which had "grown from insignificance to its present great and proud magnitude" that his integrity had been questioned. A great industrial conflict might emerge in the industry within the next few weeks and it was important, he said, that the record of the miners' leader be clear. He did not hesitate to insist that no one should have the least opportunity in the struggle to link him in their minds with any union officials who had lost the confidence of their people through connections that could not be clearly explained. Ryan took the blame, but while that closed the incident officially, it was not closed in Mitchell's mind.

The net result of all the wrangling, the attempts at compromise, the accusations and counter-accusations, was that two days before the expiration of all the agreements, the joint conferences in both the central and southwestern fields broke up without any agreement. The anthracite situation was drifting. The miners went into session alone. They had to frame some policy, if not on the duration of the suspension, at least on the method by which settlements should be made. The Ryan resolution had committed the union to a policy of national settlements. The vote of Robbins and the Pennsylvania operators, the general attitude of many operators in Illinois, Ohio, Indiana, and in the southwest since that time, suggested, however, that they were willing to settle on the miners' terms. That was a new development, and it raised the issue again sharply: Should the policy of national settlements only be followed? Or should it be modified to apply to only one of the two industries? Or were district settlements and individual offers to be considered?

At this moment the new chairman of the operators, John H. Winder of Ohio, asked President Roosevelt to secure arbitration. Mitchell, knowing how all the miners felt about arbitration since the anthracite award, wired the president to assure him that more than fifty per cent of the tonnage was willing to pay the 1903 scale and that no arbitration was necessary. Then, without any previous discussion with Mitchell as to its wording or the advisability of introducing such a resolution at this time, the president of the Illinois miners, H. C. Perry, moved that the convention authorize its national and district officials to make two-year agreements with any individual operators who would restore the 1903 scale. This meant the repudiation for the time being, no doubt, of the central competitive field, of the nine-year-old policy of the union of holding their men together until all the operators of the central field were willing to sign.

With the national suspension one day away, this resolution met with a warm response. Ryan was one of those to support it. He was sure that the Illinois operators would be forced to sign. His own resolution calling for a national suspension and national settlements had, he said, done its work by making the operators give up their demand for a reduction, thus "raising the ante," and bringing to light those men who were willing to settle on the 1903 scale. Every argument against the Perry resolution was used by Ryan for it. District abuses, such as the non-union mines owned by Robbins himself in Pennsylvania, could be more easily settled on a district basis. Although he did not say it, district advantages such as the more favorable machine differentials in two of the states, Illinois and Indiana, and the run-of-mine system in Illinois, could, the delegates felt, be better assured through district settlements than through forced arbitration. Ryan was very frank on the subject of arbitration: "I would rather go out to-morrow and fight any kind of a fight than to become a party to placing you in a position where you have to arbitrate. I never did believe very much in arbitration and I don't want to arbitrate anything unless I know I am going to win. I don't want your affairs to get into the hands of a commission that will not relieve any more of your complaints than our brothers in the anthracite fields were relieved of, and I say that with all due respect to those who handled the anthracite affairs. . . ." Ryan remembered that one of the members of the commission did not even know the meaning of the term "buddy."

Notably in opposition to this change of policy were two men from Ohio: Tom Lewis, the vice-president, and William Green, who was later to become secretary of the organization, and then, on the death of Gompers, president of the American

Federation of Labor. Were they influenced by the fact that the Ohio operators were most bitter against the return to the 1903 scale? Another severe critic was Adolph Germer, one of the most outspoken of the Illinois radicals.

Mitchell would not be a party to the closing of the discussion, and finally gave his opinion. Although he had not previously seen the Perry resolution, he believed it was possible for the union to divide and conquer. He added, "I know how popular it is sometimes to appeal to sentiment . . . in favor of a general strike, but I know too, from years of experience, how different it is when passions cool off." He spoke of the suffering in the anthracite strike; he had never forgotten the terrific burden of the miners' woes. He then offered the men their choice of his policy or his resignation. In 1902, when the anthracite convention had voted for a strike over his advice, he had made no such attempt as this to have their vote turn around him as a leader. What had happened in the interval? Here he said: "Just as soon as I fail to have the confidence and the sincere respect of the miners, not only as to my personal integrity, but also as to my judgment as to what is best for them, then I ask you men to relieve me of my responsibility." Was this an attempt to force the delegates? Was it simply a sign of his own very deep, but until that day unexpressed judgment? Did it indicate a complete lack of any desire to retain in his hands the reins of power? Was he discouraged at the lack of harmony within the administration, or was it a means of having the miners force a unity on officials who would not give it themselves? In spite of the results of a similar policy in Colorado, of the long difficulties with

West Virginia, where union and non-union men worked in adjacent valleys, did he really believe that a national settlement only at this time would mean the loss of such advance as the miners had made? Or was he overawed and lost at the magnitude of the situation before him? In the course of the next two years, there were to be hot answers to each question.

Those who thought the interstate agreement the first essential disclaimed any personal motive in the attack on the Perry resolution. Green of Ohio said: "Two years ago one of the strong reasons advanced why we should accept a reduction was in order to secure the expiration of all the contracts in the country at the same time . . . and now we are going to take another position. We cannot afford to divide our forces. The proper thing for us to do is to stand together and wait until the time when reason and moderation will possess the minds of the operators of this country." He pointed out that if Mitchell's claim, that fifty per cent of the tonnage was willing to sign the agreement now, were based on fact, it would become a matter of time only, with the continuation of the nation-wide suspension, before the remainder would sign.

For the first time in many years, Tom Lewis made a public stand on policy in diametrical opposition to Mitchell. He was fundamentally opposed, he said, to the policy of individual and sectional settlements. His twenty-two years of experience in the service of the miners, with the many national organizations which had come and gone, had proven to him that it was precisely the sectional settlement method which broke the back of the 1894 suspension. It was the omission of West Virginia in the 1897 settlement, opposed at the time by Ryan as well as by himself, which had resulted in the West Virginia problem of 1906. He pointed out the many unorganized districts, eighty thousand non-union men in Pennsylvania alone. He doubted Robbins' sincerity, asking as its test that the Pittsburgh operators sign a contract that would apply to all their mines, union and non-union. So far as the winning and maintenance of public sympathy was concerned, that was part of the task of leadership. He pointed out that the anthracite mine workers were going on suspension anyhow—or at least so he had read in the press without any direct information from his superior officer—and he urged that this could be a help to the bituminous miners. He probably knew that he was not going to win in this fight against Mitchell's policy, but he wanted to go on record this time. When the vote was taken, Mitchell won, but not before he had given notice to still another group in the union that his underlying philosophy and tactics were at variance with theirs.

He had judged the situation of the moment better than his opponents. By the middle of April the majority of the operators in the unionized fields had signed the 1903 scale. By the middle of July, the last of the die-hard operators in the union fields had submitted. West Virginia operators had made concessions even to unorganized workers. In the anthracite field, however, the only triumph was a slight nominal one: the mine workers' officials, as representatives of the three anthracite districts, attached their names to the agreement. There was no other gain.

The union was becoming something quite different from what it had been in 1898. The young leader, who had been put in because old leaders had killed themselves off with their failures and quarrels, was no longer free from sharp criticism and entangling relationships, and factions were again complicating the task of leadership. He had already stayed in power four times longer than any of the earlier presidents. The movement had grown tenfold in power and in membership, although often that had to be revived by sheer force of official effort. It had not gone as rapidly as he had hoped; perhaps not along the lines others had expected. The union was still the greatest in the country in point of size, but it was far from bringing to its men the sort of life he had in those young days hoped he could bring to the miners. It had wrested a larger share from the gilded age than any other union—but not large enough. His rise had been sudden. Some men had said it had been far too fast for his own good. Now, in the summer of 1906, morale was low. All his associates knew that Mitchell was unhappy and worried. His intimate friends knew that while the foundation of his unhappiness lay in the union situation in which he had played his part, one could not say whether the fact that John Mitchell was now drinking, was cause or effect of his unhappiness.

X

THE BREAKING MAN

MITCHELL had begun drinking after the anthracite strike, according to a friend, as a cure for his insomnia. His serious drinking did not begin, however, until 1906. As a youngster he had been called a "sissy" for refusing to drink with the boys of his own age. His abstinence was even more striking in these early days as a local union leader, for then the saloon was the cradle and the refuge of the union men, who were denied the privilege of meeting elsewhere. He had not accustomed his system to alcohol, and it developed that a very little put him under. When he did take it up, it was not beer but stronger stuff, not sociability but escape. Most of his associates drank and many of them were heavy drinkers and frank about it. It struck them as odd that Mitchell should be so concerned once he had begun drinking. His life was somewhat that of a traveling salesman in the sense that he was at home or at his office only a few weeks of each year. He needed companionship badly, but partly because he was naturally reserved, and partly also because other men were somewhat reserved with him on account of his position, it was difficult for him to make friends. He could be easy of companionship across the poker table, and that helped. He was not a man with wide interests—his work and his temperament limited him. He also needed relief from the heavy responsibilities of his task. In the early days he had found relief in the religious exaltation of his feeling of service and accomplishment in the cause. Now, in this time of disappointment and breaking harmony, he sought another type of relief and escape. The relief and escape were short, and the depression which followed a drinking spell of long duration. To him it meant a loss of self-control, a weakness, a surrender of his earlier self-dependence, for which he was not able to forgive himself. It irked him to the point of desperation when he could himself see that it interfered with the execution of his responsibility. In 1906, some operators had him drunk and playing poker for high stakes at the moment when their spokesman was secretly sending out the telegram to Roosevelt asking for arbitration. Warned by a reporter, Mitchell sobered up quickly and managed to check that move. It was the only time that he had ever been drunk on duty, but he could not forget it. Essentially he was a sensitive, introspective person, lacking a certain robust matter-of-factness about the world and about himself, feeling his responsibility to such an extent that when he saw his picture in a striker's home enshrined next to that of the Virgin Mary, he exclaimed, "My God, how can I ever fail them?" It was his attitude toward his drinking, rather than the habit itself, which was his problem. Other men, including his own people, could be tolerant of his failing; an excess of family devotion might have stood more in the way of his single-minded pursuit of his task. He could not be so tolerant with himself.

His drinking was really what he called it, an illness. He tried to cure himself. The cures were unsuccessful. He tried to keep it from the outside world as well as from his family. He began to think of it as an hereditary curse on him, something naturally beyond his control, but most of the time he was ashamed of himself.

During the year 1906 Mitchell seemed convinced that he would never get the habit under control as long as he was with the union, that a change in his living habits, a relief from the responsibility and the strain might help him in this personal problem. But in 1906 he could not retire, cheerless though the prospect was, because not even his "illness" was enough to make him willing to retire under attack.

The attack this time was not a personal one on him. No one had found him affected by his illness at any of the moments when that would have been disastrous. The miners, who practically always chose their officers on a purely personal basis, with very little regard to their statements of policies, would have been the first to resent an accusation on the floor that the personal life of a successful leader was a matter to be discussed in public. They were not puritans. Far from being a crime in their eyes, drinking meant sociability and sociability meant democracy.

The attack, when it came at the 1907 convention, was on a leader who had been willing to give up the principle—that a whole industry, no matter how scattered it was throughout the country, should be dealt with as a unit—for what he considered a more practical position. To some extent, this attack was voiced by men who had, in contrast to Mitchell's apparent opportunism, what might be called a "philosophy of unionism" to which they felt obligated to adhere. To some of these, the more vocally assertive leadership of a Debs or a Haywood, even though its results were meager in immediate increases, or in the improvement of conditions of labor, or even in the gain of union membership, might seem more successful because they believed it would lead in the long run to the spreading of the gospel of the revolutionary function of unionism. By others, men like William Green, whose fundamental concept of unionism differed very little from that of Mitchell, the objections to Mitchell's policy were made on the ground of more immediate results.

There was little way of telling at that moment to what extent the attack on Mitchell's policy screened a pure and simple personal ambition to succeed him. Later, after Tom Lewis had been in office three years, it became clear that instead of a firm policy of his own, his outstanding quality was a desire to stay in power, and to create, as Mitchell had never done, a political machine that would keep him there.

The five hundred delegates who filed into the convention hall in January of 1907 were themselves visible proof of the somber days the union was facing. Usually twelve hundred crowded the room. But now after a year of unemployment, nation-wide strike, loss of morale, probably even half of this meager delegation would have been refused places if the dues records from their locals had been strictly examined. Only in one respect did Mitchell minimize the unfavorable events of the year. By contrast with all the threats and warnings spoken by the opponents of his policy of local and district settlements, the existence of the union as such seemed a source of keen gratification. He had made another tour through the anthracite field, and found in the union only thirty-two out of every one hundred and fifty workers, and was unable to account for the cause of the apathy. To some extent he was willing to blame it on the spies who infested the whole region and "whose business it was to create discontent and dissatisfaction among the idle employees—conditions no doubt known to the operators." The I.W.W. had made some inroads on the feeling of the men toward the union. The slow financial payments from the local and district organizations to the national office were also disheartening. Only in retrospect from the inception of the union to date did some claim that the wisest policy had been pursued seem possible. The only other alternative was that the union, which now included somewhat less than half the men in the industry, would have been forced into a more disadvantageous national settlement by the prolongation of the strike. To Mitchell it seemed that in spite of the unfavorable outcome of strikes in Alabama, Tennessee and the Meyersdale district of Pennsylvania, "we have demonstrated that although an organization of peace, we have the ability to fight."

After the president's report had been concluded, the vice-president arose to give his. He did the unusual thing of launching forth, at the end of his official comment, into a criticism of the policy of 1906. He had been against it then, and he wanted vindication now. He claimed not to be making a personal fight, although there were few who believed him, but to be taking the opportunity of a convention where no agreement was to be formulated, where no plea of unity in the face of the enemy could override him, to say that the policy had already been tried and found wanting. The union had not only lost membership in the unionized fields, because the men were dissatisfied with the terms of settlement, but it had somewhat lost the power of being successful in the still unorganized areas. A great opportunity had been passed by. All the contracts had been carefully dated to end on April 1st, 1906, an arrangement for which he claimed and was conceded some credit. He painted a picture of the strike that might have been, with the central districts of Pennsylvania (Irwin and Connellsville) joining with the workers of West Virginia, portions of Ohio, the unsuccessful strikers of Alabama, and the defeated miners of Colorado in a national agreement. He raised also the question, interesting in view of the national panic that was to take place that very year, "If we are not able to establish . . . our right to organize with the present prosperity existing in this country, what can we expect when industrial depression takes place and take place it will." Then he read the record of conditions lost through the policy of sectional settlements, a long list.

His report with this attack went, together with Mitchell's report, to the Committee on Officers' Reports. Its attitude to this remarkable situation of being presented with a choice between its two main officers (and a repudiation of the policy which the delegates had adopted at the last moment when it was clear that the operators in the union field were divided), was awaited eagerly by the delegates on the floor. They knew that there were enough opponents of Mitchell on that committee as well as on the committee on resolutions to make a fight very possible. Would it be a fight?

The committee stood behind Mitchell. Rebuking the vice-president in one sentence, it spoke in the next of "the demoralized condition of our joint forces." Some of them probably meant that Lewis had been responsible for this demoralization (in view of his unwillingness to stand out against Dolan when sent as Mitchell's representative to the Pittsburgh convention). Others probably stated it as a matter of fact, without placing blame on Lewis or on Mitchell, who had a veto power over the acts of the other officers.

Then Mitchell rose to defend himself and his policy. He regretted that the press had made much of the presence of the two factions. He suggested a solution of that difficulty, if it did exist, a little more inclusive than his statement in other years about his own willingness to retire: "If, in order to maintain that unity and solidarity, it becomes necessary that the vice-president and myself step aside and take our places in the ranks, then it is your solemn duty to see that we do step aside, and it is his duty and my duty to step aside without complaint." That was not at all the kind of solution Lewis desired.

Mitchell tried to meet the attack on policy. Was it his policy alone? Practically all of the delegates had voted for it. Was it opportunism? "I have had enough experience in the labor movement to satisfy me that there is no hard-and-fast rule that can be laid down for the government of a labor union that can be applied successfully under all and every circumstance." He was not committed against national strikes in the future. Had the national strike of '97 been a success? In parts of Illinois, miners had to stay out seven or eight months, and then the settlement secured was not national; it was only a four-state settlement because only half the miners of the country had responded. In that strike, the miners had nothing to lose at its outset; in 1906 they had already much to lose. It was doubtful that the Pittsburgh district would have joined a strike in 1906 due to the situation there. In any event, not a single district agreement had been made by himself or the district officials which had not been confirmed by a convention or by a referendum. Were the locals dissatisfied with their national president? For the first time in three years, the vote cast for Mitchell as president had outrun Lewis' vote as vice-president. This he did not say so directly. He looked forward now to the reëstablishment of the interstate joint agreement, the recapture of lost membership. This time he was emphatic in asking them to place another man in his position if he could no longer serve them as they desired. He tried to explain himself to them: "They say I am conservative. Some of my friends and some of those who are not my friends say that Mitchell is too conservative. I am not conservative. I am impatient and restless for advancement. There is not a man in this country who spends more sleepless nights, worrying and fretting because we do not go forward with more rapidity. . . . It is true that I am temperate in my statements; it is true that I do not believe that the interests of the workingman are to be advanced and promoted by my standing up and calling every employer an exploiter and a robber and a thief.

"I am trying as best I can to promote a feeling of friendly business relations between the employers and ourselves; but my desire and my efforts to promote friendly reciprocal business relations do not detract one iota from my impatient desire to see the mine workers of this country among the best paid, the most humanely employed, of all the workers of this land."

Lewis and Green renewed the attack. A nation-wide suspension, they said would have saved the interstate conference. The bill of particulars covering the failures in the sectional agreements should be answered, not ignored. Other leaders rose to the defense. Things were different now and comparisons difficult. In 1897 an organizer could go through the length and breadth of West Virginia freely, but now, "about the time you leave Cincinnati going up there you would get a pretty able bodyguard to attend you." The officers of Central Pennsylvania claimed that they had gained in membership, and in conditions which included a *defacto* check-off. Dolan's successor in Pittsburgh, Feehan, was sure his district, which contained many non-union men, would have been wiped out under a national strike; the operators in his district had hoped for such a strike. Had Lewis always worked, as he claimed, for the harmony of the organization? A letter was read to the convention written by Lewis during the suspension to one of the Pittsburgh delegates criticizing the administration and putting himself on the side of the Dolan people.

There was ugly feeling about this. Mitchell took the occasion to show that he at least had not played Dolan's game, whatever it was. In a dispute about an advance for day laborers, he had declared to Mr. Robbins and another official of his company that "by the eternal they would pay that price or I would close every mine the company owned. . . . My action was repudiated and protested by the officers of the Pittsburgh district at that time." The delegates had tacked Lewis to Dolan, and he linked Dolan to something very unsavory. He had learned that the union, which had seemed to him a cause men would be willing to support with every ounce of energy in them and find some starvation a small praise for it, no longer offered that appeal. In West Virginia particularly, where every union strike had helped the non-union miners, he found out and now told the convention that the men there "will not be organized at all, strike or no strike, unless we are able to support the men in those fields from the first day they lay down their tools." Moreover in 1906 he had found that "the West Virginia operators were so much afraid of their men becoming involved in a strike . . . that they maintained a lobby in the city of Indianapolis to encourage such a strike."

The work in which Mitchell found himself had turned out in many ways to be much more encumbered with the meaner passions of men, less full of the zest, pioneer courage and selflessness which he had known in its earlier days, and which he still hoped it would regain. He knew now that although the Lewis attack was voted down on the floor, there was no real prospect of peace within the union or outside of it. The lost joint interstate agreement would involve a long and often unsuccessful struggle with the operators, and for every failure, he would meet the taunts and the "I told you so" of men who wanted him to be gone, who said it was for the good of the union that he should get out of their way.

When spring came that year, not even those who had sneered at rumors of his illness before, could deny now that he was a very sick man. In April he had two operations, one for appendicitis and one for trouble he had incurred working in the mines long ago, recurrent hernia. From the end of April until the middle of June, he lay in the hospital in Spring Valley, slowly recovering. At moments he lacked completely any belief in his own recovery. Much of the time he suffered from complete nervous exhaustion, inability to sleep. His faithful secretary, Miss Elizabeth Morris, who had taken charge of his affairs and carried on his correspondence, was writing hopefully in May that he was beginning to sleep a little. In June he felt strong enough to leave the hospital and travel west to Montana and Wyoming on organization work.

The trip West was a breath of hope. He was able to negotiate an agreement for which the miners were more than thankful. In October he was back in the hospital again, and he knew now that it was all over for him in the union. This time he stayed there a long while, nervous, sleepless, thinking of all that he might have done with his life, all that he would never do, remembering all the men whose confidence he had so wholeheartedly, seeing his successes as molehills, his failures as mountains, tired, eager for peace which seemed impossible in this world, unable to bring himself for a long time to the point where he wished to recover. But people needed him. His wife and children were not alone in letting him know that. Messages of appreciation and condolence were coming in all over the country. He was reminded that he was loved and trusted, that no matter whether he had to leave the union or not, he would live with them always. In the face of so much reassurance he slowly gathered strength to believe that he would recover. It was the turning point.

Again he tried to get back to work, attending a preliminary conference with operators' representatives from the four central states in Indianapolis, and going to New York. But at Christmas time he was in the hospital at La Salle, near Spring Valley, because of a second relapse. As he was speaking before the conference on the 20th of December, he was suddenly stopped in his speech by acute pain, which turned out, however, to be only a bursting of the adhesions. Believing that he was dying, he said: "I believe my time is at hand. I do not care. I am half glad that it is almost over. If I do not see the men again tell them for me that whatever success I have brought for them has not been for pecuniary benefit. It has been because I love them." They took him back to his hotel, and there he reached out and sought for some peace which he was unable to give himself. He asked his friend Ryan to send for a priest and to be allowed to join the church of his family and many of his friends. He did not die. He lay again in the hospital at La Salle, discouraged with himself. Outside the hospital room, the work of the union went on. He was unable to attend the various district conferences and conventions. In October he had told the miners that he could not again be a candidate for the office he had held ten years. Now they were balloting for other men. W. B. Wilson was running against Tom Lewis; his friend Ryan was running for secretary. His withdrawal had not lessened the factionalism. It was not heartening to think about the union now when another agreement had to be made.

There were some things, of course, about which he could think with less mixed emotion. The district agreements and joint conferences still gave the miners the terms they would have obtained under an interstate conference. The reports as to the growth of the organization were again optimistic in all except the anthracite fields. In his ten years of service, the real earnings, as well as the rate of payment of the miners, had increased far more than in any other unionized or non-unionized industry. The cut of the working day from ten to eight hours in the bituminous field, from ten to nine hours in the anthracite field, were gains that could not be eaten up by the increasing cost of living. Miners all over the country were freer to live and to buy where they wanted to; much of the industrial serfdom had been done away with. The miners like Mitchell, theorized little about "industrial democracy" but could oppose some very effective controls of their own over measurement and discipline to the sovereignty of the operators. In the union districts death rates from accident were lower than in the other districts. The union had broken the tradition of defeat and kept it broken. He had taken his part in awakening the workers, not only to a sense of the injustice of their position but to a feeling that within their ranks they held the power to abolish that injustice. He had helped break down the barriers of racial prejudice and craft jealousy. His union had been outstanding in a successful answer to the challenge of the machine; the machine had been accepted but in such a way that the workers and their union were not menaced. Unemployment and low annual earnings, due to excess production in the industry, to inefficient and superfluous mines, had not been eliminated, but the percentage of unemployment and over-expansion was lower in his day than ever before or after, except during the World War. The day when it became a test of the union's ability to eliminate over-expansion and inefficiency lay fourteen years ahead. He would have welcomed the findings of technicians and analysts but none existed, and he was busy union building in the days when the very existence of a union was a miracle. Of all the industries in the country, the coal mining industry was, even after the disasters of 1904 and 1906, most strongly organized.

When the convention opened, in the middle of January, he was able to appear again and looked better, somewhat stouter and healthier than before he had gone to the hospital for the third time. It was a Mitchell convention from the first day. An Illinois delegate presented him with a loving-cup. Later a delegate from Montana and Wyoming said he did not have a silver loving-cup, but "we came with love in our hearts . . . and $2,700 in money, the check for which I now present to him." The simple speech brought tears to the eyes of many delegates. Mitchell, who did not know what he was going to do to earn his living once his tenure of office terminated, was touched. He said: "I don't know what to do—I don't want the money. If I could ask you to take it back and give it to the men who sent it here, and feel sure that they would know I appreciate their confidence as much as if they had given me as many millions as they have hundreds, I would feel better about it. I know how kindly their feelings must have been when they sent it here. I am sure I do not know what ought to be done, but I shall ask either that the money be returned to the men, and my thanks given to them for it, or that it be given to my wife who has probably made more sacrifices than I.

"The ten years, or nearly eleven years, I have been away from home have been a period of alternating joys and sorrows. When the men did well, when they got something more than they had, I was happy and exuberant, and when the men lost something they sought or yielded something they had, it made me sad and depressed. During all those times I tried to maintain as impassive a countenance and appearance as it was possible for a man to do; and yet, much as I feel that I have suffered myself, I know that at my home there has been greater sacrifice made. . . . I have been away most of the time, seeing them two or three times a year, never more than three or four days, except when I was in the hospital.

"I prefer that you take the money back, but if you cannot do that without danger of giving offense to the people who sent it to me, then send it to Mrs. Mitchell and let her use it to educate our boys."

The miners proceeded to business. Mitchell read his report. Membership had increased, but not enough to make up for the losses of the previous year. Trade union sentiment was strong and the "I.W.W. and other malevolent agencies" had not been able to "pervert the minds of the mine workers." In the far west, in Washington, Montana and Wyoming, there had been a phenomenal growth; excellent agreements for the miners had been secured. The fate of the interstate movement in the central fields had not yet been determined. Mitchell did not mention a rumor that there would be no meeting until after he had gone, though a conference had been set for January 30th. He spoke, as usual, on the necessity for observing contracts. He set before them the solemn figures on accidents and deaths in the mining industry and urged them to put forward more strenuous efforts in the matter of new legislation and in the enforcement of existing legislation. The employment of boys in the bituminous fields of Pennsylvania was still a problem. The critical matter of determining the wages and conditions of the machine miner, who more and more was replacing the pick miner, was before them.

It was perhaps in the spirit of leaving behind him a constructive memorial for the old, the wounded and the helpless, that he proceed to propose an insurance plan for accident and death benefits along the lines "followed so successfully by the railroad brotherhoods." He was glad to report that relationships with the Western Federation of Miners were improving, though there was much still to be done on conflicting provisions in the laws of that organization. He stated his opposition emphatically to a proposal for the consolidation of the two unions.

Then he gave what can only be termed a valedictory. He reviewed the phenomenal growth of the membership, the improvement in the wages and conditions of the mine workers. For these he did not claim credit, he took only the pleasure of partnership.

His relations with the district and sub-district officers, with few exceptions, had been harmonious. Indeed, "to have continued in the service of the miners and to have contributed further to the amelioration of their working conditions" would have been an "honor and a delight." But his health would not permit it. He pointed out the grave responsibilities of office and asked for his successor—and he did not as yet know whether it would be Wilson or Lewis—the loyal cooperation of the organization. He thanked his colleagues and prayed "that wisdom may guide your pathway in the future and that God's blessing may rest upon our movement."

After the other reports from Lewis and Wilson, came a flood of resolutions: one from a Texan delegate suggesting two years' salary; another from Kentucky that an educational department be founded in the union with Mitchell at its head; another from a Kansas group that he be retained as special adviser to the organization; finally, one which provided six months' salary until he had recovered his health. Mitchell asked the rejection of all of these, but the last. Of this, with its expression of appreciation, he said it would be kept "as the proudest and most honored legacy of my life." He felt that his own happiness and contentment justified him in declining to accept any reward in the shape of remuneration; the organization had paid him from year to year all that he would accept from them. He would even refuse the money in this last resolution, which would give him an opportunity to travel and have medical attendance were it not for one fact: "It has been charged—and unfortunately it is true—that in the years of the past the champions of the workingman have received their reward in kicks and blows. Scattered over our country . . . lie the bones of men, gone and forgotten, who rendered to our great cause as much service and made greater sacrifices than I have done.

"I wish it were possible for the spirit of the great John Siney as well as the spirits of Dan McLaughlin and those champions who fought and gave their lives without reward, who went to their graves—hungry in some cases, starved in others—not with the love of the people they unselfishly served. They had to die and a long time passed before they got the honor they deserved while they lived. And were it not that this stain might be removed from workingmen's organizations of our country, I should ask you to defeat the resolution you have now under consideration."

The delegates might show their appreciation of him by a rising vote while these gifts were made, by tumultuous applause, but the convention was not to adjourn before Mitchell was to be faced from the same platform with a sharp challenge of everything he stood for and almost everything he had accomplished. "Big Bill" Haywood—still secretary of the Western Federation of Miners for a few months before he was let go because of his continued affiliation with the I.W.W. after his union, which had once sponsored that body, now condemned it—had just come out of prison, in connection with the accusation against Moyer and Pettibone and himself for a part in the murder of Governor Steunenberg of Idaho. Where Mitchell was of middle height or less, dark, a quiet speaker, a conservative leader, sentimental, dramatic in a somber way, Haywood was burly and big, over six feet high and heavy, a gigantic Cyclops, melodramatic and picturesque.

Both were union leaders, leaders of miners. The contrast in their physical appearance was no greater than in their philosophy. One came from a sick-bed, the other from an imprisonment considered unjust. One was the leader of a successful industrial union which had restricted itself to unionism, which believed in law and order and the sacredness of contract. The other was the leader of a revolutionary, rather unsuccessful union which had participated in the socialist movement, in the founding of the I.W.W., only now to depart from that path for the one of "business unionism." The coal miners and their leaders, much as they may have disagreed with the three accused men—incensed as some of them were at the rôle played by the Western Federation with regard to their union—rallied to the defense of Moyer, Haywood and Pettibone, and were the largest contributors to the defense fund as well as to the public spirit aroused in their behalf.

Haywood's speech was a long indictment. He recounted the history and vicissitudes of his union in a rambling, often vague, but always picturesque fashion, taking random shots as he passed at capital, at the government, at the sham called "law and order." He reminded the coal miners of their woes in common with the metalliferous miners, telling them of the 1903 strike in Colorado, which happened to be a sore point with Mitchell: "I recall one instance when eighty of them (coal miners) were gathered together and driven like cattle for twenty miles. Some of them fell prostrate on the way. Others were beaten as they traveled, beaten up by Cossacks on horseback who were driving them. These are your brothers! When they arrived at Trinidad they were driven up to a horse trough and watered like cattle. . . ." It was a bitter, powerful speech of the resentment of the outlawed and the oppressed.

He pleaded to the coal miners for a different state of affairs, when labor, united in one organization, would resist its exploiters. Specifically he pleaded for consolidation of the two organizations so that "it will be impossible for the smelting companies to operate their smelters because the coal miners will not mine coal to be used as scab smelters. The railroads will see the necessity of refusing to fire their engines with coal mined by scab coal mines, and the general result will be the organization of labor extending over the industries of this nation." It was the dream of the nation-wide sympathetic strike of all industries. He appealed to the miners whose convention, he said, was the "most important gathering without any exception that meets in the United States" which "can wield a greater influence for weal or woe than any other organization . . . be it labor, religious or political, to consider all these questions "from the workingman's standpoint."

He went on: "Do not enter into agreements that are going to bind you up for a period or make it impossible for you to do anything else for yourself or for anybody else. Leave your hands free so that you can fight whenever a fight becomes necessary." He asked them to include a clause in their contracts, "as the coal contractor does," freeing them of the contract in case of strikes: "If it is wrong for an individual to enter into an agreement, and likewise wrong for the local union to enter into an agreement, I say that it is wrong for an international to enter into an agreement unless that international takes into consideration the interest of the working class. . . ."

When he sat down, he had challenged three things sacred to Mitchell; respect for contracts one had signed, the unity of all labor regardless of industry, and the respect of law and order. He had said, furthermore, something which Mitchell did not touch on: that unless the oppressed of the world had a battle cry, and a great, almost unobtainable objective, they could never move forward at all. All this called for an answer. Mitchell did not go outside the field he had learned to know. He reminded the miners of the time when they had no contracts. "Our contracts are not perfect, our system is not ideal, but there is not a man whose memory runs back ten short years who would go back to the system then prevailing if he could. . . . We had not the contract and we had only half as much wages. I say only half, and we worked longer hours." Contracts might not represent an ideal state of affairs, but he would think cautiously and measure well before he returned to the time when he could strike as he wanted to—and the operators could strike back as they wanted to. These were the times when miners' wives went from city to city, begging bread for men at work."

Now that he was leaving, he felt himself free to say something a little stronger than he had ever done before: "I am going to make one reservation. . . . I believe there is a time which comes in the history of peoples and nations when treaties should be abrogated . . . when men should rebel . . . overthrow their government and change the whole system under which they live; and I believe there are conditions of industry that make sympathetic strikes not only proper but highly advisable." If the time should come when, in his opinion, a sympathetic strike would aid both the coal and metalliferous miners, he would advise it. Did he forget he was retiring? He ended on the note of hope for the continuation of the friendly relationships that had replaced the time when he and Haywood had "misunderstood each other." So far as a consolidation of the two organizations was concerned, he had expressed his views clearly before. An industrial union in the coal industry should include all coal miners, anthracite as well as bituminous, precisely because anthracite coal competed often with bituminous coal. The industrial union followed the market and not the craft, and therefore could not include miners of silver, gold and copper. He believed however a plan acceptable to both sides could be worked out.

On the final day of the convention, the election results were announced. Lewis had won by a two thousand majority out of almost one hundred and forty thousand votes. John P. White of Iowa, who was later to become president, was elected vice-president in the only uncontested office. Ryan had decisively beaten John Fahy for the office of national secretary-treasurer. Wilson, who may have been beaten because of the fact that he had already been elected as a member from Pennsylvania to Congress and would have been forced to hold down two jobs when Mitchell had collapsed from the strain of one, was asked by Mitchell to present the President-elect to the convention. In later years, it was charged that there had been fraud in the counting of votes. Lewis addressed the convention insisting again that he, Mitchell and Wilson had never had any personal differences, and shook hands with them in proof of his statement. White and Ryan spoke, and then Mitchell made the final oration, wishing his successors good luck: "If there be one man in the union who imagines that I would feel gratification at the failure of my successor, I hope he will dispel from his mind that false and infamous idea. . . ."

Mitchell concluded: "And now . . . may I ask that if, unconsciously or inadvertently I have in my life done aught to wound or injure any man, he will give me his forgiveness, just as now, as I pass from the limelight into the shadow, I forgive any man who may have done wrong to me. I ask you to favor me by demanding, if any man charge me with wrong when I am no longer here to speak in my defense, that he face me."

He had not said his last farewell. His office still had three months to run. During that time there were vain attempts to revive the interstate conference. When that failed, a special convention was again called to consider the policy for the future. It was at this time that the *Indianapolis News* commented that Mitchell was "still the masterful and foreseeing general" the wisdom of whose Sick Man of the Bosphorus policy in 1906 was justified and proven by Lewis' advocacy of it in the 1908 situation.

The time came for the last farewell. On March 19th, 1908, the delegates passed him in file and said good-by, tears streaming down their cheeks. "God bless you, John," they said, and Mitchell answered: "Hold up the hands of your officers. A pat on the back, a word of good will does more good than all the criticism you may give him. . . . I say to you now, gentlemen, may God bless our movement. May he look down on it with that favor, with that grace He has given it so far as I have been connected with it. Gentlemen, you are going back to our people. Encourage them to carry out this policy . . . even though in detail it may not meet your views as a whole.

"My boys—and I will call you my boys even though some of you are nearly twice as old as I am, I regard you as my boys—be as good union men now and in the time to come, as you have been while I was directing your affairs. God bless you again! I am leaving this movement, not because I want to leave you, but because I have to leave you."

There is a story that this last speech was made after the convention had already adjourned, with the singing of *America.* Mitchell had left some time before, in low spirits, and had gone to a nearby barroom. The failure to revive the joint movement, Lewis' succession to the presidency, his illness, the fact that he was completely at loose ends as to his own future, the strain of the whole proceedings had unnerved him and he sat alone, drinking. He was not quite sober when Ryan arrived with the news that the delegates had refused to leave until they expressed to him once more their feelings of appreciation. They felt the greatest of their leaders was going, and they wanted to hear from him once more. He mounted the platform in tears himself as he brought tears, by the sentimental, but moving speech, to the eyes of his people. His life—or the best part of it—was leaving him. He was not ashamed of his tears.

XI

THE PROMISE OF HIS YOUTH

WHAT would John Mitchell do now? He was only thirty-eight years old, and in a sense, his life was still before him. He had the responsibility of a large family; of six children born, four were still living, all of them needed protection, the education he had not had himself. In order to take care of them, his first concern had to be the regaining of his health. A rest was made possible by the six months' salary allowance voted him by the convention. But after that?

He could not go back to his craft, to the job of coal mining, because of his hernia. There was also a widespread feeling that the labor movement needed him in some way, if not as an officer, as an editor, as founder of a labor school, or as ambassador to the general public. He had acquired knowledge and gained experience and public confidence that neither he nor his friends in the labor movement wanted to have lost. Could not the American Federation of Labor utilize a man of his training and ability? There were few, if any, who were his superior. Then, too, he had for some time lived on a social and financial scale approximating that of the small business men of the country, and it seemed natural to him to continue the march upward. He had a keen sense of having neglected his family in spite of the fact that they may have enjoyed the bread that miners' children could not eat. Even that was endangered now, for he had no savings out of his salary, which had been $1,200 up to 1903, and $3,000 thereafter.

He was a broken, or almost broken man, and his passive disposition might have led him to be happy in a quiet retirement were financial cares not so pressing. His family, too, was not ambitious for him in a worldly way, though his friends might be.

What had happened to other presidents of the miners' union which might guide his choice? During the six months after he had stepped down from his office, and while he went from one cure to another, this question gave him something to think about. John B. Rae had built up a small business for himself. Phil Penna had become commissioner for the coal operators' association in Indiana. John McBride had, after one year as president of the American Federation of Labor, become labor reporter for an Ohio paper. Michael Ratchford had retired for an appointment on the United States Industrial Commission and had later become Industrial Commissioner for the state of Ohio. Lesser officials had had similar experiences—either they had become commissioners for the operators or had held some political office. On the whole, most of the miners seemed to bear comparatively little bitterness toward their leaders for entering such positions. This was, of course, only in those instances in which the commissionership was in districts where collective agreements with the operators existed, and when their former leader in his new office gave them what they considered a square deal. They realized that in such cases his knowledge of their methods and aims might even be to their advantage. He did not for a moment think of following Penna before him, or setting a precedent for Tom Lewis after him, by going across the line and becoming an agent for the operators. Not only was he still second vice-president of the American Federation of Labor, but he carried with him into his retirement a great sense of the dignity of the office he had held for ten years and a pride in the name he had kept for himself. Whenever any union leader did cross the line there were, and Mitchell knew it, always some who would sneer that he was now where he had belonged all the time. Not only that, but men like Penna were expected to earn their salaries—which meant that they were forced, whether they liked it or not, into opposing many of the aims of organized labor. The miners might not always see the conflict as sharply as a sensitive man put into such a false position would see it and be troubled by it.

The other possibility was a political appointment. The miners, who often played a great deal with local and state politics, had some feeling of awe for the national government. Appointments by the President of the United States, even when given for obvious political purposes, were held in high respect. Mitchell had been spoken of for Congressman or Governor of Illinois in 1902 and 1906. At one time, in 1904, before the Democrats had decided they could oppose what seemed to be the radicalism of Roosevelt with the conservatism of Parker, Mitchell had been mentioned as a very logical nominee for the vice-presidency. Election to Congress had been achieved by one of the anthracite leaders and by the secretary of the national organization, W. B. Wilson, who later became Secretary of Labor under President Wilson.

Mitchell's district might well have elected him. There also were rumors that he might be appointed United States Commissioner of Labor to succeed Carroll D. Wright. There were now rumors that President Roosevelt, who was known to admire and who had already appointed him to an honorary post in connection with the award to him of the Nobel Peace Prize, would make him member of a commission to supervise the labor problems connected with the building of the Panama Canal. In March the newspapers interviewed him concerning a rumored nomination for Governor of Illinois.

Then, in May, a boom was begun for him as vice-presidential nominee on the Democratic ticket. His old friend, John Loftus, in an interview in Scranton, stated that while Mitchell would not make an effort to secure the nomination, he might consider it were it preferred. It was generally known that Mitchell was not a regularly enrolled Democrat. No definite action was taken until July. His name was brought up by an Illinois delegate to the Democratic national convention and it was said he had the backing of New York. But Mitchell let it be known that he did not wish his name considered.

Samuel Gompers, who years later was to suggest W. B. Wilson as the official nominee of the Federation for the position of Secretary of Labor when President Wilson himself had Mitchell in mind, and who still later was to help Mitchell in an appointment on the New York State Industrial Commission, made a comment on this affair in his autobiography which was not without its stinging quality: "Rumor persisted that the desire of a large number of delegates to the convention was to nominate John Mitchell. . . . His deeply sentimental nature had caught the fancy of a large number of people. He had emerged from the successful leadership not only of the bituminous but of the anthracite miners' strike. I am sure that Mr. Mitchell did nothing to encourage or discourage the movement, and of course my associates and I and the other members of the Executive Council could take no action to further his nomination. I have fully explained why Mr. Mitchell's name was not mentioned in connection with the nomination in the convention, but it was rumored that because Mr. Mitchell was so great an admirer of President Roosevelt and under such obligations to him, he could not become known definitely as a member of or an adherent of the Democratic Party. However, for some reason that has never been explained, Mr. Mitchell's name was not presented to the convention."

Nothing came of any of these rumors. President Roosevelt was retiring from office, and Taft, who had different sympathies, was being nominated to succeed him. In August, Mitchell solved his difficulty by accepting the position of Chairman of the Trade Agreement Department of the National Civic Federation. This was a department particularly created to promote the growth of collective bargaining between employers and unions of workers. In this work John Mitchell believed that he could help to bring about a better understanding of the aims and methods of trade unionism on the part of employers. He was aware, of course, of the opposition to the Federation by a group within his own union. The position paid well, $8,000 a year. Moreover, it required no extensive traveling and, for the first time in ten years, Mitchell hoped to be together with his children, of whom he had seen so little and for whom he was passionately ambitious. He was as conventional in his desires for them as any other good father. He wanted them to have comfort, a good education, the companionship of both parents, and a far easier life than he had himself known.

This was not exactly, as many thought, an evasive way of crossing the line to the employers' side. Many of the big industrialists were members of the Federation and not all of them followed anything approaching a union policy in their dealings with their labor. On the other hand, Samuel Gompers, James Duncan, first vice-president of the Federation, Mitchell himself and many other labor men, were also members. Where some of the labor leaders and editors might distrust Mitchell's judgment of men and their motives, and consider the Civic Federation simply a luxurious emasculation of the unwary, the presence of Gompers, who always managed to convey the impression that he knew very well what he was about, disarmed some of the comment on Mitchell's choice of a livelihood. Hanna had been an important member of it and had apparently tried to make his union views prevail there. He remembered that during the anthracite strikes the Civic Federation had helped to secure some settlement at a time it was needed badly. It had also helped secure recognition for traction workers in New York City in its early career and had aided other organizations. He had never in his life been averse to gaining for labor any of the things that could be gained for it by showing the employers that labor leaders were reasonable, decent human beings, even though others might doubt the relevancy of such an appeal.

For the next two years his life was comparatively quiet. Perhaps the change in his mode of life, his confirmation in the Catholic Church, had helped him at last to overcome the habit. According to friends, until a period shortly before his death, for some ten years, Mitchell was free of what he thought his particular cross in life.

The work was a let-down; no big things were given him to do. He had a suspicion that he was being sidetracked. He wrote Gompers, himself a member of the Civic Federation, on August 9, 1910, giving instances in which his work, through his influence with employers and with banking interests, had brought results. This was really the work of the conciliation department, which was concerned with the settlement of strikes. "To be perfectly candid I am not entirely satisfied with the service I have been able to render. In the well-organized trades the representatives of the union and the employers in interest have been able to work out their own problems, and it is much better of course that they should do so than that there should be an intervention on the part of outside forces; but in the partially organized trades in which unions are not strong and in which the officers are not thoroughly experienced, I am handicapped because employers in these industries usually contend for the open shop and it is well known among employers that I am in favor of the union shop."

Because of his very prominence as a union leader he had

often been rejected as a go-between. When his services were requested, it was in cases of strikes which were practically lost.

In spite of its advantages in domestic comfort and peace, the position with the Federation was a spiritual decline for this man who was still young. He was a member of his union and a delegate from his local in Spring Valley to miners' conventions. From the moment Lewis took office, the official journal of the union began to reprint newspaper items which carried insinuations that "the gentleman in New York" would hinder the new administration, and for that reason Mitchell was very careful. Present at a convention of miners in the anthracite fields in October, 1908, he was introduced by Lewis with the statement that the two had always been on friendly terms. The stenographer who took notes of the speech, tells of the meeting: "Mr. Mitchell stood alongside, gravely regarding him, then turned his somber face to the audience. 'You have just heard what Mr. Lewis said. I want to take this opportunity, I want to say right here in the presence of this gentleman, in the presence of you men here who are most vitally concerned, that this organization can never be built up by tearing me down.' This was said so earnestly, so dramatically, so triumphantly that the miners jumped to their feet and shouted. 'This organization cannot be destroyed by any one man. It is not I, or Tom Nicholls, John Fahy or Tom Lewis who builds this organization and keeps it going. It is the man at home, the man who is never heard of, the man whose picture is never seen.

" 'When I was a young man, I used to think that if I could

be President of the United Mine Workers, I'd use all my power to put the employer where he belonged. The time came when I was President of the United Mine Workers, when I had all the power possible, not given me by any laws but by the affection and confidence you men reposed in me.' Again the impassioned, triumphant voice brought the miners to their feet.

" 'For ten years I was president but I did not revolutionize the industrial world. I did not solve the labor problem. It will never be solved. We only hope to gain little by little more and more of our proper share of what we produce. The men coming after you will not solve it. After us will come other John Mitchells, Tom Nicholls and John Fahys who will help you along, but who cannot accomplish much unless you give them your support. Be loyal to them as you were to me.' Coming down from the platform, Mr. Mitchell was surrounded by the miners, many of them kissing his hand in their wild enthusiasm, several Italians embracing him as they exclaimed, 'Oh, Johnny de Mitch.' "

The test of the sincerity of these men's loyalty to Mitchell and of his own loyalty to the miners' union came two years later, at the 1911 convention of the union.

In the intervening years, there was much done to "tear down" both Mitchell and Lewis. Violent objection to Lewis from many sources resulted in the conversion of the *Mine Workers' Journal* into a campaign sheet for him for forthcoming elections. It was charged that Lewis kept the larger portion of the organizing force in unionized districts, as a power to squelch any opposition to him. When Socialist leaders, Duncan McDonald amongst them, for whose vote Lewis had made bids previously, opposed him, they were discharged. W. D. Ryan resigned his office as national secretary-treasurer. John Walker in 1909, and William Green, Lewis' former ally in 1910, almost defeated him for the presidency, and would have done so, it was charged, if not for fraud in the counting of the votes. In this opposition to him, Lewis insinuated Mitchell had a hand. Lewis did not accept with grace the fact that Mitchell was still retained as second vice-president of the American Federation of Labor. The reason given for this continued election was that Mitchell was now under sentence in the famous Buck Stove and Range injunction case. Mitchell, as presiding officer of the 1908 miners' convention, which had passed a resolution asking for the boycott of the Buck product, had been held in contempt of an injunction already issued against such boycott. He had received a sentence of nine months in prison, Gompers one year, and Frank Morrison, Secretary of the American Federation of Labor, for six months. Appeal was taken, but meanwhile, for the sake of public standing, Mitchell was reëlected as second vice-president. Another reason might have been that the Executive Council mistrusted Lewis and believed he would not last more than one term in the miners' union.

Apparently Mitchell himself gave no cause for Lewis' feeling until 1910. Late that year a special convention was held in Indianapolis to fight out a serious revolt of the Illinois district against Lewis. The Illinois officers to a man charged that Lewis and certain members of the National Executive Board had made an agreement for the district, without consulting the district membership or officers, on terms far below what the district would have accepted. When a referendum was forced by protest, the operators, it was said, sent agents throughout Illinois to persuade the men in their locals to accept these terms. The referendum was against Lewis and he refused to have it counted. A special convention was called, at which Mitchell was present as a delegate from Spring Valley. He declined to speak, despite wild applause from the delegates followed by Lewis' invitation. He said he was a delegate only, without any preconceived notions of the situation.

At this convention, there was much talk of having a new president, to which Lewis retorted, ". . . You'll not have a new president." Delegates shouted, "Let someone else do some talking." The conventions of the past two years had been free-for-all fights. At one point, when the fighting almost broke up the session, it was Mitchell's plea for order which permitted it to go on.

Finally, on the seventh day, Mitchell, by request, made his statement. He explained that he had doubted the wisdom of his coming and that only the seriousness of the crisis made him decide to do so. He commended the national officers but lined up with Illinois in its objection to having a scale made by a National Executive Board on which no single Illinois member was represented. (Frank J. Hayes, the Illinois representative, who was later to become president, had on election to the vice-presidency been told summarily that that office was no longer at national headquarters and his function as head of the organization work abolished.) Mitchell referred back to the history of the organization to show that no action comparable to that of this National Executive Board had ever been taken. Lewis' action was repudiated and the district negotiated its own agreement. Mitchell's action lost him, for a good many years, the friendship of Harry N. Taylor, who was hit by the action of the Illinois miners. Taylor had complained in past years that Mitchell leaned over too far backward to prove that the friendship did not influence his judgment. Now Mitchell was out of office, why must he use his influence in this way? Mitchell was amazed and hurt at Taylor's attitude. He did not forget it, for he was attached to this large, jolly, burly man who could make him laugh. He had done his duty, that was all.

In the 1911 elections Lewis was finally defeated for the national presidency and announced his intention of going back to the "picks," giving an unsolicited promise never to sell his services to the other side.

But Lewis, as presiding officer over the 1911 convention, had his revenge. When the convention met in Columbus, a telegram was received from John Mitchell, saying that he could not be present as delegate. It was necessary for him to be in Washington in connection with the arguments in the Buck Stove and Range injunction proceedings before the Supreme Court.

After election returns were in, a bombshell was exploded in the report of the Committee on Resolutions. Under the heading of "National Civic Federation" it considered several resolutions, some condemning that body and asking for approval of that condemnation to be carried to the American Federation of Labor, as several international union presidents were members of the former body. One resolution, signed by several men from the anthracite district, made no reference to action by the American Federation and provided merely for the immediate expulsion of John Mitchell, who was the only member of the miners' union holding membership in the Federation. The report of the Committee on Resolutions accepted the latter resolution.

A minority report was signed, protesting on the ground that the Civic Federation was a voluntary organization, separate and apart from the labor movement, acting only in an advisory capacity, although admittedly it "had not fulfilled the expectations of friends and members of organized labor." However, it was said, since John Mitchell was the only member of the union connected with the Federation, and since he was at the moment sentenced to jail because of his services to the labor movement, the action of the Majority Report would be construed as an attack on him rather than on the Civic Federation.

For two days thereafter, the convention discussed the course of procedure; some delegates argued that it was not the function of the Resolutions Committee but of the Constitutional Committee to recommend such changes; others charged that it was a personal plot against Mitchell. The condemnation of the Civic Federation was not a new issue. As far back as 1904, some of the more radical delegates had brought in resolutions condemning it, but they were reported on unfavorably by the Resolutions Committee of the time. In 1911, however, the convention delegates were bankrupt, bitter and defeated. They had listened to an account of a failure of a strike in Northern

Colorado, in Nova Scotia, in the Irwin fields, and elsewhere; they had been burdened with assessments.

On the first day of the convention Mother Jones in her speech had made a bitter attack on the type of men representing the "public" in the Civic Federation, and on the purposes and aims of the Federation, which she said were to chloroform any real revolt or radical change in the labor movement. It is pretty safe to say that in 1911, although the Civic Federation was not exclusively concerned with attacks on Socialism as it is to-day, it was no longer influential in bringing together employers and unions in important situations as it had been at the outset.

Strangely enough, in 1911, the old leaders of the radical wing were opposed to the report of the Resolutions Committee. Duncan McDonald arose to protest that since only one member of the union was involved, he should be present to be given a chance to defend himself. The answer was given that when the saloon keepers and mine bosses were excluded, they were given no such opportunity. McDonald felt furthermore as did Walker, that the matter was one to be brought before the American Federation of Labor, since only one delegate of the Mine Workers was involved, but many officers of the American Federation of Labor were active on the Civic Federation. He felt the issue should be one of the Civic Federation and not of a single man. He insisted on the point, against the statement of Lewis as chairman of the convention, that it was not an issue of Mitchell but of the Federation.

Some delegates argued that the large vote cast for John Mitchell that year as delegate to the American Federation of

Labor—he had received 113,285 votes, the next highest being Walker with 85,887½, Hayes with 77,175, John P. White with 75,982, and Lewis with only 69,503—showed the real feeling of the miners for Mitchell. The answer made to this was that had the miners known the real nature of the Federation and Mitchell's connection with it, they would not have cast such a heavy vote for him. Still others who spoke for the resolution said that John Mitchell was too valuable a man to be sitting in idle banquets with Belmont and Carnegie and other enemies of organized labor—that he ought to be on the lecture platform, acquainting the country with the aims and ideals of organized labor. Lewis himself spoke in favor of the resolution and one of its movers stated that he drew it at the request of the "administration."

The report of the committee finally was left undecided because one of the signers of the resolution was accused of having been a strike breaker.

On the eleventh day, this attempt of the Resolutions Committee having failed, the Committee on Constitutional Changes added to a resolution providing for the withdrawal of a member when he became a mine boss, a similar provision for a member of the National Civic Federation. This gave Mitchell the opportunity, at least, of making a choice. On this resolution 443 votes were cast for, and 378 votes against. This vote was not accepted, and another vote was taken the next day on roll call, which resulted in 1,213 votes for the report and 967 against.

Some of the delegates had forgotten the applause with which they had greeted Mitchell's request, in 1908, that if he were ever accused, he would be given an opportunity to defend himself.

Word was sent to Mitchell, now in New York, by friends, of the action of the convention. Mitchell's son recalls the family anxiety at this time, with Mitchell under sentence in the Buck Stove and Range case, with his own organization against him, and the well-ordered little world he had established for his family at Mount Vernon threatened with ruin. The family opinion seemed to be that he should not resign his post with the Civic Federation.

What was a man to do? His growing family, his new scale of living, his inner conviction that even if he were doing no great amount of work for the Civic Federation, it was not bad work in the sense that it harmed labor. The opinion of many of his friends, all led him to be inclined to keep his office. What was there against it? The vote, large in fact, but which would obviously never have been taken if Lewis had not been interested in securing revenge for Mitchell's part in the repudiation Illinois had administered to him, was that of a union he had left as officer. True, he was still its representative on the executive council of the American Federation of Labor. True, he still faithfully kept his union card in the Spring Valley local, but he also knew though other men feared it was not so that he had no desire to return to the fighting involved in any attempt to resume leadership in the union. The large vote against him proved not only what he knew already, that men demanded almost impossible success from their leaders and remembered that success but a short time. The delegates of the anthracite districts where he had been hailed and acclaimed as nowhere else had been almost unanimously against him. He could take what comfort there was out of the fact that Illinois voted four to one for him. Why not get out of the union? Did it have anything more to give him?

The feeling that now he was earning a salary, but that the best days of his life, his highest dreams and hopes had been and—for his own sake—must always be attached to the union, brought him to a swift decision. His standing with those men was more to him than success now could be. He was trying to remember himself as a young man, before he had been beaten, trying to put away obstacles between himself and that picture, eager to see himself still within the dynasty of his early promise, eager to have others see him there, too.

He resigned his position with the Civic Federation at once. He wired the delegates that he submitted to their wishes but would "live in the consciousness that the men and women at home for whom I worked so many years will not concur in your conclusion." That remained his hope—that the quick judgment passed upon him without hearing could only be the result of political manipulation, that the men at home not caught up in it would be more willing to do him the honor of believing that he would have accepted no dishonorable employment. He was somewhat heartened at the many letters of protest and requests for a referendum sent in by locals throughout the country, and the long discussion of a protest movement in the anthracite region. All this came to nothing, partly because he did not lead any fight for his own vindication, partly because his resignation from the Civic Federation had settled the matter in one way, and partly because men really needed some whipping-boy for their piled-up defeats and were willing to let the Civic Federation feel some of the lash.

Samuel Gompers suggested that he make a fight on the forces in the union back of the resolution, and offered him the columns of the *American Federationist* for "a service to the cause of labor and justice, for which no like opportunity might present itself in a decade." Mitchell said that he could not write the story without reflecting upon some of the men in the miners' union, and preferred to do that within the realm of the union rather than through an outside paper. But when, after Lewis was out, he did write three articles for the *Mine Workers' Journal*, it was not the kind of fight Gompers expected. No old scars were opened up. He did not set himself as a target for retaliation, as a ball for the muddy feet of men. When he finally wrote, it was only a denial of the claim that the Civic Federation had ulterior purposes that tended to chloroform labor into helplessness and an outline of its helpful activities. In his letter of resignation to the Civic Federation, he regretted that it had been a subject of gratuitous attack and said that it had stood consistently as an advocate of righteous peace.

Mitchell's choice was probably the one which a sincere trade unionist could make under the circumstances, despite its loss of a position which now paid $10,000. It is difficult to judge what might have occurred had he stayed with the Civic Federation, because of the change in the nature of its activities shortly afterwards. An official of the organization commented unfavorably on Mitchell's lack of initiative in the attack against what that official considered the Socialist menace. Somewhat later, in this connection, when Mitchell was on a lecture tour and was asked to insert in a Milwaukee address a few words of attack on Socialism, he refused to do so even though it might have meant a cancellation of the lecture.

Once again then, in 1911, at the age of forty-one Mitchell was footloose and casting about for a position in which he might, in his way, be of service to labor. His very caliber and standing made the creation of any post within the miners' union or the American Federation of Labor difficult, even if he would have wished it. He was not a man who could be made a subordinate, and few leaders were willing to retire in his favor.

XII

THE RESIGNATION TO SUCCESS

THE dilemma which Mitchell faced in 1911 was perhaps the same one which confronts all ex-presidents. He faced it younger than most; he was only forty-one. All his experience, instead of preparing him for the best years of activity, had hampered him. The miners' cause was the situation he knew most intimately. Lewis was out but he knew that no miners' president could appoint him, even if he should wish to accept a humbler post, to a position where his very presence might seem to challenge the administration. There seemed to be no place for him in the American Federation of Labor outside of the honorary post he held. Other unions had their own leaders. There was very little opportunity for a trade union leader to function unless he did so within the unions themselves. What about the balance of the labor movement in its more radical, political phases? He was unacceptable to the radicals, and he was no radical. He was confronted with the difficulty which had in his day also confronted many men outside of the labor movement who saw in it the cause for hopeful change and yet had found no way of working with it.

Mitchell, through the initiative and management of Miss Morris, spent the next two years lecturing to the public at large. In these lectures, he was still the man who had told the anthracite miners in 1908 that he had not solved the labor problem and that it would never be solved, but that the union was their only safeguard against a return to the days of greater suffering. He had too long been a responsible leader to become a crusader now, a man with some new flaming vision or word with which to stir others. He had learned then that a responsible leader, with wage negotiations always in the offing, cannot say reckless, even if glorious things, because his own people would then judge him by his own avowed aspirations or condemn him for a radicalism greater than their own, and because the employers were always eager to seize on the pretext of radicalism. All this had made crusading a luxury or even a vice.

When finally he was free from responsibility, both to the miners and to the Civic Federation which had employed him, his lectures were dignified statements of the moderate and humane aims of unionism, its ameliorative social function. It was what he believed. If it was not exactly a rousing call to a new order, neither was it what was later to be called business unionism. It was a program of progressive modification of the existing order, and not a glorious gesture to be remembered and to rally one on to the overthrow of that order. His did not stir the occasional listener who might have responded with enthusiasm to the evangelism of a Debs or an Emma Goldman. To the large masses of the common people who flocked to the Chautauquas, or the average student who heard him at the universities, he brought a revision of the then common belief that the labor movement was a purse-stealing, black-jacking, bomb-throwing raid on American institutions, but made them understand it as an essential to their own welfare as well as to that of its membership. To many throughout the country to whom labor was anathema, who would not admit Gompers into the circle of respectability until after he had proved his patriotism in full, but who could be roused to enthusiasm by Roosevelt's rally at Armageddon, he brought the simple statement that the progress of the human world was made by small but continual gains, and this, in spite of his own somewhat romantic personality.

While he had originally been conservative because he had been very rapidly and suddenly thrust into situations which made him aware of their limitations and of his own, he remained conservative because, like most of the miners of the day, he accepted the lingering American traditions. His neighbors and fellow members in the union had no philosophy which radically challenged these. He had been a small-town man and he remained sensitive to the opinion of his neighbors. While he could say immediately after the grueling hearings of the anthracite commission that "the average wage earner has decided that he must remain a wage earner," in his own life he had seen evidence of the ability to rise to success. Although he knew he was something of an exception, he did not want to feel himself alienated from other miners, to see himself as an isolated exception. He saw none of the irreconcilable conflict which the I.W.W. were preaching. He wanted to believe in peace because he was essentially hopeful that strong union organization carried with it progress. The people from whom he had come found in their neighbor's eyes a respectability through peace, ostracism through violence. To this, although he was doubtless unconscious of it, was added the fact that many of the people he admired were successful, wealthy, such as Hanna and other men in the Civic Federation, who wanted industrial peace badly and believed it possible, without any great change in the character of the labor movement. His essential conservatism had not been modified by the history of radical efforts. He had seen the Socialist Labor Party and the I.W.W. rent by factionalism, unsuccessful in organizing any but the migratory workers of the West, flaring up in the Lawrence strike of 1912 only to leave the workers without any permanent organization. He had seen tired radicals, and liberals for whom unionism was not sufficiently progressive, turn from one solution to another, only to dissipate themselves in theoretical and factional disputes.

During these two years he asked himself again the question that constantly haunted him—to what extent, judging from the developments since his retirement, had he failed the people who trusted him and failed himself?

He might have drawn some comfort from what he saw. After the slump which had lasted from 1906 till 1911, the miners were again going ahead. His old opponent Lewis had been defeated and later had gone over to the operators' side in the non-union fields of West Virginia, taking with him office records which the operators later used. In that year the anthracite districts were again rebuilt and the miners were given the long-needed machinery of mine committees to protect them. In 1913 membership gained over one hundred and ten thousand in spite of losses in Cabin Creek and other portions of West Virginia. The Ludlow massacre in Colorado was still a year in the offing. Could this be failure? He knew that while in the soft-coal fields he was one of many leaders, in the anthracite district he remained to many "the greatest labor leader the world had ever known"—peacemaker at home, a valiant warrior against the enemy. He knew that while for years men had cursed the awards of the anthracite commission, they now had the mechanism of local pit committees by which they could protect themselves. While he had been in office, there had been ever-present the danger that the non-union fields—notably West Virginia—would break the union hold. That danger remained and was growing. He had never tried to do there what he had tried in the anthracite field. There had been over-development of the industry. There still was. To some extent the efficient and high cost mines might have been closed down or kept from opening up by a uniform wage scale throughout the central competitive field. When he entered office, labor was bearing the entire burden for inequalities in costs of production so that every operator could sell his coal in the market. Instead of forcing upon the central competitive areas a wage scale which would have eliminated all of the over-development of the industry, the union had found it necessary in 1898 to compromise on the "principle of competitive equality," which had resulted in the elimination of many mines and since then he had been forced to divide his time and attention to protecting it from low-cost competition of non-union mines in the outlying districts. He had given up the interstate agreement, and eight years were to elapse before it could be won back. Twenty years after he left office, the over-development of the industry was to result in giving up the agreement in the central competitive fields for the second time. The over-production that was to come in part through machine mining still lay in the future. Its great influence on the union was a matter to be charged up to the leader succeeding him. Meanwhile he did have to his credit a record of higher wages than those obtained in other industries, of lower hours obtained, and a record of negotiations honorably conducted. It had become, with the building trades, the backbone of the American Federation of Labor. His influence on American trade unionism, through his success with collective bargaining, had been inestimable.

It was a not unimpressive record. While developments and changes in the industry were to make some of his omissions and failures seem of great importance later on than they did now, he could truly go before the country, not only with sincerity and almost unchallenged integrity, but with a record that very few other men—if any—who had taken up the cause of the weak could boast.

In these four years, Mitchell himself had developed intellectually. In 1903, with the very substantial collaboration of Walter Weyl, he had written *Organized Labor* which, though it was a record of accomplishment for a period in which labor had had few historians and fewer analysts, was too guarded and factual an account to reflect the man himself. In 1913, he published a second volume *The Wage Earner and His Problems*. It showed a careful study and an understanding of labor legislation. It revealed a more questioning attitude toward the fundamental righteousness of the political and economic order of the day. His belief in American institutions did not blind him to the value of continental experiments along other lines. The vogue of the Taylor system provoked a careful, though not technical analysis, of the wastes of management and of capital.

During this time he retained his connection with the executive council of the American Federation of Labor. It was the one possible bridge back to the career which had made him a name. He retained that office only because he had been indicted with other A. F. of L. officials on the charge of boycotting the Buck Stove and Range Company, and the accident that the Federation leaders did not seem to want Lewis to take his place on the council, and White had refused to accept an eighth vice-presidency. Could something be made of this chance?

At this moment, a great opportunity came his way. Early in 1913 the newly elected president, Woodrow Wilson, was choosing a man for the position of Secretary of Labor. It was known that he had John Mitchell in mind. A little pressure by Mitchell's friends, a word spoken by the Federation leaders, would doubtless have confirmed him in his choice, for he saw no one else on the horizon of Mitchell's caliber, and supposed that the labor leaders of the country would agree with him. There was no one who brought this pressure to bear, and Mitchell still held to his feeling that, even where there was competition for honors and offices, they must come to him unsolicited. It was then that the American Federation of Labor, apparently unaware of President Wilson's intention, and unapproached by Mitchell himself or his friends on the subject, recommended another man.

This was a blow. It was true that he had several ideas about unionism which conflicted with those held by Samuel Gompers and others in the Federation. On one of these, the necessity of industrial unionism in the mines, he had fought and won. Gompers had seen him withdraw from the struggle to become vice-presidential nominee on the Democratic ticket in 1908. He had seen him withdraw from the Civic Federation without making the sort of bitter attack on his opponents which Gompers had recommended to him and for which he had offered Mitchell the columns of the labor paper. He had been irritated by Mitchell's objections to his public statements at the time of the first decision in the Buck Stove and Range case. Mitchell then had suggested many modifications in Gompers' language and had lectured him to the effect that "Dignity and proper restraint are at this time as essential as courage and candor."

To Gompers all this seemed confusion between the dignity of labor and its virile militancy. Himself a man with a robust, uninhibited enjoyment of life, a shrewd political ability, a sharp verbal resourcefulness, and a live hatred for his opponents both in the employing and in the Socialist ranks, Mitchell's somber reflectiveness seemed provincialism; his self-control and tolerance seemed to Gompers more fitting for an impartial chairman than for a man who might, it was hoped, really fight effectively labor's battles in the cabinet. He passed over Mitchell and recommended—which made the preference more striking—another miners' leader, W. B. Wilson. He had been secretary of the union while Mitchell was president. It could be said in his behalf that he was an enrolled Democrat who had had Congressional experience. Even after that recommendation had gone in, it would have been possible for Mitchell to have had his friends bring pressure to bear on Gompers to recommend him as well. This he did not do. He went further. He begged his friends to support Wilson.

A few months later, at the Seattle Convention in 1913, a movement was set on foot by many in the American Federation of Labor to have him succeed Gompers, who was then sixty-three years old. There was a large group which felt that Mitchell, as the younger man, had not only greater potential energy, but was more acceptable to the public. Had Gompers accepted the proposition that he should retire and be given the post of honorary president and the editorship of the *American Federationist*, Mitchell would have been willing to run. But Gompers did not feel himself to be an old man and refused to retire; he finished his life in the saddle. Mitchell would not oppose him, nor was he any man for Mitchell, who had always tried to avoid the exigencies of a struggle for power, to challenge. Mitchell felt that his own dignity and that of the Federation was at stake. Perhaps if there had been some issue which he saw as vital to the welfare of unionism in the country, he might have pocketed this sense of dignity or pride. The issue against craft unionism did not present itself to him as a challenge. Years before, after he had successfully fought the battle of industrial unionism for the miners, he told John Walker who wished to introduce a resolution for its introduction in other industries, that "we are too young to advise." It would certainly have been a hopeless issue to go with before an organization that was founded on the sovereignty of these crafts. Although he did not feel as bitterly opposed to the Socialist Party as his associates on the executive council—he made a distinction between it and the other two political parties—still he was outspoken in his belief that the labor movement was not to be made the tail of any political "kite." He had no fundamental fight with the Federation on its political policy.

That was the closing of a chapter. From now on, though he might stay on the fringes of the labor movement, the bridges that might have carried him back were down. He could not go on lecturing indefinitely, for lecturing at its best was a pause in the life of a man of action during which he tries to bring people up to the place he has reached. Perhaps at this moment he might have gone into business. He was casting around still at loose ends, when Governor Sulzer of New York State appointed him Commissioner of Labor, after the Senate had twice vetoed his appointment. Even that went out from under his feet, for Governor Sulzer was shortly thereafter impeached and removed and the appointment nullified. But now friends were stirring and in March, 1914, Governor Glynn, Sulzer's successor, chose Mitchell for a four-year term as member of the Workmen's Compensation Commission. In June of 1915 he was transferred and appointed Chairman of the Industrial Commission of New York State, a post he held till he died.

It was a position which drew largely upon his ability to see both sides of each case, to secure the slow progress possible through legislation. He took particular interest in the cases dealing with compensation to workers for industrial accidents, and often, instead of delegating to deputies the task of holding the individual hearings, attended them himself. This type of legislation was then new in New York State. His especial interest in its fair enactment had its permanent effects not only in New York but in other states. He thought many times of the great number of his fellow-craftsmen dead and crippled each year in the great toll of life taken by the mining industry.

His new work also involved a certain amount of conciliation of industrial disputes. He brought to them the same essential appeal of his personality which had marked his conduct as a labor leader, the appeal of a decent man who expected to find the same quality in others. During one hot summer there was a strike in an industrial town in northern New York. It had been going on for some time, and there seemed to be no possibility of an amicable settlement, the employers steadfastly refusing to meet with a committee of the workmen. The employers, most of whom had formed into an association during the strike, represented the respectable, staid, long-established element in the community, controlling not only the industrial but the financial life of the city. The workers were despised foreigners, underpaid and overworked. Continued refusals to meet with the strikers had, it was rumored, aroused a desperate spirit in them and there was talk of dynamite being stored in town. Unable to make any headway, Mitchell's associate left for Albany with a threat of an investigation. It was August and the members of the Commission were due for a vacation, but instead they were summoned to come to the town to make the required investigation. His associate told the story:

"The day of the proceedings was hot and sultry. Everyone

was irritated and tense, as worker after worker was called and gave the same desperate story, and employer after employer refused, on advice of counsel, to give his version of the strike. Mr. Mitchell, slow, deliberate, was in the chair and he seemed to be the only one who could keep his temper. Even those commissioners who were supposed to be friendly to the manufacturers' side had become incensed at the attitude of the employers.

"Finally one of the younger employers was called to the stand. I sent word to Mr. Mitchell that this was the only man in the employers' group in whom I had any hope. Mr. Mitchell proceeded with the regular questions. What was the man's name, his firm's name, how long had he been in business, the number of his employees, etc.

"When he came to the question: 'Will you tell us what you know of this strike?' he received the same irritating answer: 'On advice of counsel I have nothing to say.' The case seemed hopeless.

"At this moment, Mr. Mitchell asked the witness, 'Are you a member of the same association as Mr. —— (an employer who had gone to Canada to avoid examination)?' He received an answer in the affirmative. Then Mr. Mitchell, slowly, dramatically, deliberately and earnestly said, 'I am going now to perform a most unpleasant duty but one which I believe necessary under the circumstances.'

"Then he read into the record a letter from Mr. ——, the absent employer, obscene and vituperative, in response to a request from a committee of his workers to meet with them. The whole matter took but a few moments, but it seemed longer because of Mr. Mitchell's low, deliberate reading. The air was very tense in that swelteringly hot room and there was hardly a sound until the witness cried out: 'My God, I can't stand it any longer. I've always been respected in the community, respected by my employees. I've lived here all my life and I'm sick to death of all this.'

"I leaned forward and urged Mr. Mitchell to take advantage of the opening made. But to the surprise and dismay of the other commissioners, Mr. Mitchell adjourned the hearing at the moment for the next day without further explanations. Next morning there came a committee from the employers to meet with the workers' committee and eventually a settlement was made.

"It was a dramatic episode—spontaneously dramatic. Only a good man, himself clean, would have known how to appeal to the decency of the man on the witness stand."

Into work of this type he settled with some half-satisfaction. It was good work, useful work, there could be no challenge of the kind that had accompanied his work with the Civic Federation. He settled down into it and grew stout. His family was growing up in his Mount Vernon home, his children getting some of the education he himself had needed badly. Now he was out of the clamor and strain of the labor movement, passing, in the name of the public, judgments on the social value of its aims. He himself was part of the public which Hanna had said was either employer or employee. To remind himself, to keep himself from forgetting all that had once been close pain to him, and was now perhaps in danger of being lost among the papers of legality and the dust of office routine, he made a point of going to the East Side of New York City in order—by seeing the swarming, dirty streets, the crowded, sour-smelling tenements—not to lose the consciousness of poverty and misery and unemployment. Once a week it was his church.

His associates were mainly fellow officials and some few union leaders. His social contacts with the four hundred were very slight. He remained a lonely man. At heart his interests were far away. He once said: "I have good friends there, I like them, they like me. But we have so little in common. They know nothing of mining. You can't know what it means to fall in with some old mining friends and talk over old days." Whenever old acquaintances from the mine fields came to New York, he was insistent that they should visit him, and for some of them, these visits became pilgrimages to Mount Vernon. Occasionally he managed to get down to Scranton and see his friend, John Loftus, and feel the warm beat of approval that came up to him from the mining districts on Mitchell Day. He decided that when he died it was there that he wanted to be buried.

When the war came, and other labor leaders were being called in on the determination of national labor policy, he continued to serve rather as a representative of the public. He found scope for his organizing abilities as chairman of the State Food Commission, of the Federal Food Board for New York, of the State Council of Farms and Markets. From a distance, he could see the miners' union grow, its membership prosper, the A. F. of L., whose conventions he still attended, grow to a membership of five million. Labor acquired a recognized status in the country.

During these years and especially those before America entered the war, he had found a new vocation. In his early career he had been distressed by the loss of all his savings, two thousand dollars, in a bank failure. He had known insecurity of employment throughout his youth, and twice after his success. In the conversation of the men with whom he had associated at the Civic Federation, there had been frequent talk of the market. He had begun to invest his savings on some of the tips. The market became to him a puzzle like life that could be solved, a puzzle like his own life which had to be solved. He began studying it carefully. It was the poker he had always liked, but on a large scale, magnificent, sometimes appalling scale. It became his drink. In his earlier years he had been willing to gamble on the number of cars passing a given corner, on races between raindrops, on the color of horses appearing on the street. To some extent it was sheer excitement, to some extent an attempt to beat a game. About his own success in life he had his opinions. But the stock market might be a game where he could win. It was a curious passion for a man who had no love for luxury, who practically never used the automobile he purchased for his family in the last years of his life, who had never been concerned in his days as union official with the size of his salary. During the war, there were many industries which profited, whose stocks rose sharply. Mitchell was acquiring wealth. But the war which brought him prosperity was the last blow for this weary man. He was patriotic enough, his sons all served in the war, but his patriotism lacked the belligerent joyousness of other men. He could believe with Wilson that it was a war for democracy, but death and combat saddened him. He did not even feel useful any longer, not even necessary to his family who were now well provided for. His term of office would soon draw to a close and he could retire. Then what? In his weary despair, he turned again to drinking and to the self-disgust which accompanied it. During the war there were many industries which profited, whose stocks rose sharply.

When the war came to an end in France, the miners tried to free themselves from their obligations not to strike during the war and to obtain wages that would allow them to keep up with the high cost of living, and their officers met the situation by saying, "We cannot strike against the government," and men in his home state of Illinois retorted, in a special convention, that the war had been over a long time and that they had been sold out, Mitchell was taken to a New York City hospital for an operation. It was not an important operation. Nobody worried about it. It seemed a success. But after it was over, pneumonia set in, and on September 10, 1919, he died. He was only forty-nine years old. It was the day on which Calvin Coolidge, Governor of Massachusetts, settled the Boston police strike.

The hearty body, the prosperous man, had at last been successfully betrayed by the weary, gentle eyes which had seen enough and grown tired eleven years before they finally closed. In the last week of his life he showed the same lack of desire to live that had characterized his illness twelve years before, in 1907.

While many had been caught up in the industrial developments of the war, in the changing status of labor, had seen the officers of the railroad brotherhoods sitting with a stopwatch in the gallery of Congress, threatening a national railroad strike if the eight-hour law were not passed, had seen revolutionary unionism in Russia, unionism compromising a revolution in Germany and new figures and faces taking the place of older ones, Mitchell's death was a reminder of days past, of a struggle eclipsed in the changing structure of the world in which labor was expected to have a great, if not dominant part. That was 1919.

Others saw him as a man who had served supremely well according to his abilities. Louis F. Post wrote in *The Public:* "John Mitchell is dead. His body is laid at rest in Scranton and the crowds of simple-minded folk who thronged the cathedral are ample testimony that John Mitchell lives. The value of men like Mitchell to the nation lies in their ability to communicate the spirit of democracy to others. It transcends any mere material contribution to society.

"A newspaper in commenting upon John Mitchell's career compares his services to society favorably with those of a captain of industry preceding him by a few days. Somehow the spirit instinctively rebels from such a comparison. After all, what services has any duke of steel or prince of anthracite rendered which entitles him to the thanks of the world. The most that can be said of him is that he has striven for a great prize and has obtained it. The fact that he has attained eminence may be a proof of ability. But ability is a common enough virtue. We smother enough of it annually in the slums to make hundreds of captains of industry. . . . Not so with Mitchell. . . .

"John Mitchell came out from the coal mine as Lincoln came from the cabin. John Mitchell somehow won through these adverse artificial circumstances. While he was still alive his voice was always lifted to make it easier for others, to break through the hard crust that shuts out opportunity. We can honor his memory best by making it easier for many to serve as he served."

To the surprise of many it developed, when his will was offered for probate, far from being a poor man, he left an estate of almost $250,000. For a moment, the first eulogies written of his death were forgotten among those doubts which always surround money of labor leaders: Had it been acquired honestly?

His estate consisted of: Bonds, $71,900; Notes, $2,266.66; Stocks $156,000; Bank Deposits $14,129.99. The bonds and stocks were held in such corporations as the Armour Company, the Baltimore and Ohio Railroad, the Chicago Rock Island and Pacific Railroad, and the New York Central Railroad. Curiously enough, in the last speech Mitchell ever made —before the state convention of the New York State Federation of Labor—for the first time in his life, he had committed himself to government ownership of the railroads. A later accounting showed an increase in the value of his estate to $347,151. It listed shares of the Texas Pacific Coal and Oil Company, and eighty-seven shares of the Peabody Coal Company of Illinois, the latter valued at $21,897.90. The will, later declared invalid on some legal technicality but the provisions of which were carried out, provided for equal division between his wife and children of all but $10,000, of which $5,000 was to go to Miss Morris, and the other $5,000 to be distributed between two half-brothers and a cousin.

To those who had all along regarded Mitchell as being too conservative, the very existence of this fortune was in itself all the proof they wanted of the fact that John Mitchell, as president of the miners, had not been honest. Others wondered how such a large fortune had been accumulated in so short a time.

According to intimates of Mitchell, when he left the union in 1908, he was a poor man. He had lost his savings of $2,000 in 1905 and his illness had eaten up any possible further accumulation. The six months' salary given him by the miners, the fund of the western miners turned over to his wife, and $3,500 which represented the proceeds from the sale of his Spring Valley house, were his total assets—altogether about $7,500. His salary at the National Civic Federation was $8,000. His family lived modestly in Mount Vernon and Mrs. Mitchell was slow to give up the housekeeping habits of a miner's wife, doing her own housework for many years and living well within his salary.

As far back as 1903, Mitchell had begun investing in stocks with the first proceeds from his book and from his articles. He continued investing with the savings from his Civic Federation salary. He had not made very profitable investments, however, until 1911. With his resignation from the Civic Federation, he had begun his study of the stock market. Frequent turnovers as a result of close study of the market, and the use of values during the war pyramided his holdings, it is said. On the whole he kept clear of the mining industry or of anthracite-carrying railroads. The stock in the Peabody Coal Company, a corporation in the organized fields of Illinois, according to accounts, was given him in exchange for the stock of a non-mining company which was absorbed by the Peabody Company as a subsidiary. To show his scrupulousness in this respect, friends relate that when he took office during the war with the Food Commission, he sold his holdings in the Sheffield and Borden Milk Companies. These are the explanations, so far as they are obtainable.

Mitchell was, in these years after he left the active leadership of the miners' union, interested in the accumulation of wealth as most men are, primarily for the sake of his family. He did not profess to be a radical or to consider the investment in stocks and money-making evil in themselves. His union had invested its funds in stocks and bonds and had thus increased its financial strength. It is doubtful that he would plead for himself that the individual is helplessly and intricately bound up with the whole capitalistic system. He had, as far back as 1903, said that in his opinion it was no crime to be rich. He had said with equal emphasis that his leadership had no basis of pecuniary ambition.

In every sense, this quarter of a million was a tragic waste. He had no capacity for enjoying wealth; it was in itself to him a mark of the decline of his career. Mrs. Mitchell died shortly after he did, and one wonders whether the four children derived sufficient advantage in education and comfort from the money to compensate for the doubts it cast on their father. In his success in the financial field, his success in the union field is often forgotten, especially by men outside the labor movement who knew little of it and so were willing to suspect the worst.

There has been no proof given, either directly or in the retrospective interpretation of Mitchell's leadership, that any act of his had contemplated remuneration from the employing class after he had retired. The miners stood by him. Their confidence was unshaken. They had seen no act of Mitchell's which had given them cause for suspicion. They usually were stirred to no vehement indignation when their leaders withdrew to the employers' side. Even this Mitchell had not done, and as soon as he had been accused of it, he had met the test by giving up ten thousand dollars a year in return for the honor of remaining a member of their union. They had no objection to men of wealth. They wanted to believe that men from their ranks could succeed. It pleased many of them to know that not only when he had been with them, but after he had left them and become an official representative of the miners to the nation, he was a success.

Five years later they thronged from all over the country to unveil a monument which they paid for by contributions in a public square of Scranton, the city of his greatest triumph and his final place of rest. Powderly had been mayor in the eighties, but Mitchell was remembered as the great tribune and the first American labor leader to be honored by such an unanimous throng and such a monument.

On the monument is his statue, and the statue does not look as he did. All through the coal valley there live men and women who have eaten the bread and tasted the freedom he fought for with their fathers. In their minds they keep a warm memory of him—and that memory may be a myth of what the man was. In the marble of the monument is graven, "Champion of Labor—Defender of Human Rights," and it stands there for all to see in the center of the city. A mile away he lies under a stone as unpretentious as himself. He had tried to do what he could for the men and women with whom he had worked and suffered. He had not always succeeded. But there had been moments in all his weariness when he knew and could be happy in the thought that his work and his memory would make it more possible for others to bring about for his people a fuller and richer life.

BIBLIOGRAPHY

A. BOOKS AND ARTICLES BY JOHN MITCHELL

BOOKS

"Organized Labor; Its Problems, Purposes and Ideals," by John Mitchell, *American Book and Bible House,* Philadelphia, 1903.

"The Wage Earner and His Problems," *P. S. Risdale,* Washington, D. C., 1913.

ARTICLES

Autobiographical article, *The Circle,* June, 1909, Vol. 5, p. 330.

"Recognition of the Union," *The Independent,* Vol. 53, 1895, Aug. 15, 1901.

Articles on travels abroad, included in *United Mine Workers Convention Proceedings,* 1905.

B. MAGAZINE ARTICLES ON JOHN MITCHELL

Morris, E. C., "John Mitchell, the Leader and the Man," *The Independent,* December 25, 1902.

Steffens, Lincoln: "Labor Leader of Today," *McClure's,* August, 1902.

Warne, F. J.: "John Mitchell, the Labor Leader and the Man," *Review of Reviews,* Nov., 1902.

Weyl, W. E.: "The Man the Miners Trust," *The Outlook,* March, 1906.

Weir, Hugh: "John Mitchell," *Human Life,* October, 1908.

—— "Mitchell the Man Who Kept His Head," *Current Literature,* April, 1912.

—— "Appreciation," *The Public,* Sept. 20, 1919.

C. PUBLICATIONS OF INDUSTRY

UNION

Proceedings of Annual Conventions of United Mine Workers, 1898-1911.

United Mine Workers' Journal, 1891-1925.

The American Federationist, 1891-1925.

Proceedings of American Federation of Labor.

PUBLICATIONS OF TRADE

Coal Trade Journal, weekly, New York.

Fuel, Chicago.

Black Diamond, weekly founded 1886, Chicago.

Coal Trade Bulletin, Pittsburgh, Pa.

JOINT PROCEEDINGS

Joint Interstate Conferences of Coal Miners and Operators of Illinois, Indiana, Ohio and Pennsylvania, 1899-1908.

D. GOVERNMENT AND OFFICIAL PUBLICATIONS

UNITED STATES

Report of Commission on Industrial Relations, Washington, 1916.

Report of U. S. Industrial Commission, 1901, Vol. 12, Report on Mining.

Investigation of Wages and Living Conditions in Coal Mining, Hearings on H. R. 354, 64th Congress, 2nd session, 1917.

Investigation of Wages and Working Conditions in the Coal Mining Industry, Hearings on H. R. 11022 before Committee on Labor of H. R. 67th Congress, 2nd session, 1922, Testimony of P. H. Penna.

Report on the Anthracite Coal Strike of 1902, U. S. Anthracite Coal Commission, 1903.
Hearings before Anthracite Coal Commission of 1902, Testimony, 1902-1903.
Bulletin No. 43 of the Bureau of Labor, Report of Carroll D. Wright to the President on the Anthracite Coal Situation, June 20, 1902.
Report on the Labor Troubles in the Anthracite Regions, 1887-88, 50th Congress, 2nd session, H. R. No. 4147.
Coal in 1919, 1920 and 1921, by F. G. Tryon and Sidney A. Hale, Report of U. S. Geological Survey, Part II.

STATE

A Report on Labor Disturbances in Colorado, 1880-1904 (58th Congress, 3rd session Senate Document 122).
Statistics of Coal in Illinois, Supplemental Reports of State Bureau of Labor Statistics, 1881-1905.
Report of Special Committee on Northern Illinois Coal Strike of 1889.

E. GENERAL WORKS

Beard, Charles and Mary: "Rise of American Civilization," *Macmillan Co.,* 1927.
Bishop, Joseph Bucklin: "Theodore Roosevelt and His Time," New York, *Scribner's,* 1920.
Bruere, Robert W.: "The Coming of Coal," New York, *The Association Press,* 1922.
Buchanan, Joseph: "Story of a Labor Agitator," New York, *The Outlook Company,* 1903.
Commons, John R.: "Trade Unionism and Labor Problems," Boston, *Ginn and Company,* 1905 edition.
Croly, Herbert: "Marcus Alonzo Hanna; His Life and Work," New York, *Macmillan Co.,* 1923.
Evans, Chris: "History of the United Mine Workers of America," 1920.
Gompers, Samuel: "Seventy Years of Life and Labor," New York, *E. P. Dutton and Company,* 1925.
Goodrich, Carter: "The Miner's Freedom," Boston, *Marshall Jones,* 1925.
Hamilton and Wright: "The Case of Bituminous Coal," Institute of Economics Studies, New York, *Macmillan Co.,* 1925.
Hoxie, Robert F.: "Trade Unionism in the United States," New York, *D. Appleton and Co.,* 1917.
Hourwich, Isaac: "Immigration and Labor," New York, *B. W. Huebsch,* Revised edition, 1922.
Jones, Mrs. Mary: "Autobiography of Mother Jones," Chicago, *Charles H. Kerr Co.,* 1925.
Justi, Herman: "Organization of the Employing Class and Other Papers."
Korson, George H., ed.: "Songs and Ballads of the Anthracite Miner," New York, 1927.
Leiserson, William M.: "Adjusting Immigrant and Industry," New York, *Harper and Brothers,* 1924.
Lubin, Isidor: "Miners' Wages and Cost of Coal," Institute of Economics Studies, *McGraw-Hill Co.,* 1924.
Lloyd, Caro: "Henry Demarest Lloyd, 1847-1903," New York, *G. P. Putnam's Sons,* 1912.
Lloyd, Henry D.: "The Strike of Millionaires against Miners" (The Story of Spring Valley), Chicago, *Belford, Clarke Co.,* 1890.
—— "The Lords of Industry," New York, *G. P. Putnam's Sons,* 1910.
—— "Men, the Workers," compiled by Anne Withington and Caroline Stallbohm, *Doubleday, Page & Co.,* 1909.

McDougall, Walt: "This is the Life," New York, *Alfred A. Knopf, Inc.*, 1926.
McNeill, George E., ed.: "The Labor Movement, the Problem of Today," 1886.
Merriam, C. E.: "Introduction to H. Gosnell's 'Boss Platt and the New York Machine,'" *University of Chicago Press*, 1924.
Powderly, T. V.: "Thirty Years of Labor," Columbus, Ohio, 1890.
Perlman, Selig: "A History of Trade Unionism in the United States," New York, *Macmillan Co.*, 1922.
Randell, Robert: "John Mitchell Exposed," *New York Labor News Co.*, 1905.
Roy, Andrew: "History of the Coal Miners of the United States," Columbus, Ohio, 1907.
Roberts, P.: "Anthracite Coal Communities," New York, *Macmillan Co.*, 1904.
—— "The Anthracite Coal Industry," New York, *Macmillan Co.*, 1901.
Raushenbush, H. S.: "The Anthracite Question," *H. W. Wilson Co.*, New York, 1924.
Roosevelt, Theodore: "An Autobiography," New York, *Charles Scribner's Sons*, 1921.
Savage, Marion D.: "Industrial Unionism in the United States," New York, *Ronald Press*, 1922.
Suffern, Arthur E.: "Conciliation and Arbitration in the Coal Industry of America," *Houghton Mifflin Co.*, 1915.
—— "The Coal Miners' Struggle for Industrial Status," Institute of Economics Studies, *Macmillan Co.*, 1926.
Sullivan, Mark: "Our Times, The Turn of the Century," New York, *Charles Scribner's Sons*, 1926.
Wallas, Graham: "Human Nature in Politics," New York, *Alfred A. Knopf, Inc.*, 1921.
Warne, Frank J.: "The Coal Mine Workers," New York, *Longmans, Green & Co.*, 1905.
Welbourne, E.: "Miners of Northumberland and Durham," Cambridge, England, 1923.
Wolman, Leo: "Growth of American Trade Unions, 1880-1923," *National Bureau of Economic Research*, 1924.

F. MAGAZINES AND PERIODICALS

"The Weekly People," organ of *Socialist Labor Party*, 1901-1908.
"Social Democratic Herald," *Social Democratic Party*, 1901-1908.
"Miners' Journal," organ of *Western Federation of Miners*, 1901-1908.

CONVENTION AND JOINT PROCEEDING COMMENTS

Indianapolis News, 1900-1908.
Pittsburgh Press, 1899.
Columbus Dispatch, 1898.

STRIKES OF 1900 AND 1902

Chicago Record Herald.
Philadelphia North American.
New York Times.
New York World.
The Outlook.
Review of Reviews.
The Independent.
Digest of Hearings before the Anthracite Strike Commission, reprinted by Scranton Tribune, 1903.
Scranton Republican, May 30 and 31, 1924.
Coal Issue of The Survey Graphic, New York, April, 1922.

INDEX